Out of Place

MICHAEL HOUGH

Out of Place

RESTORING IDENTITY

TO THE REGIONAL

LANDSCAPE

Yale University Press

New Haven & London

Published with assistance from the foundation
established in memory of Amasa Stone Mather
of the Class of 1907, Yale College.

Designed by Richard Hendel.
Set in Ehrhardt type by G & S Typesetters,
Austin, Texas. Printed in the United States
of America by Halliday Lithograph, West
Hanover, Massachusetts.

The paper in this book meets the guidelines
for permanence and durability of the
Committee on Production Guidelines for
Book Longevity of the Council on Library
Resources.
10 9 8 7 6 5 4 3

Library of Congress Cataloging-in-Publication
Data

Hough, Michael.
  Out of place : restoring identity to the
regional landscape / Michael Hough.
    p. cm.
    Bibliography: p.
    Includes index.
    ISBN 0−300−04510−7 (cloth)
          0−300−05223−5 (pbk.)
    1. Landscape assessment.   2. Geographical
perception.   3. Landscape architecture.
4. Regional planning.   I. Title.
GF90.H68   1990
712—dc20   89−35127
                        CIP

TO BRIDGET

Contents

Acknowledgments

Many friends and colleagues have helped in the making of this book and I am deeply grateful to them all, for without their support and encouragement it could not have been written. Some need special mention. I owe much to Judy Metro, senior editor at Yale University Press, for her help, encouragement, and professional skill in making the book what it has become. She and Judith Calvert, the manuscript editor, expertly disciplined its form and turned my often over-enthusiastic prose into something readable. Particular thanks go to the Canada Council for the grant which allowed me to travel and spend time discussing the issues that have shaped the book with the many people I met in universities and practices across the United States and Canada.

I am grateful to my good friend Suzanne Barrett who read early drafts of the manuscript and gave me much needed advice; to Simon Miles for numerous, and often enlightening discussions on environmental issues; to Grady Clay for his incisive comments and writings on how to read the landscape; to Robert Newbury for his penetrating views on place and the environment; to Randolph Hester for his incisive editorial comments on my chapter on tourism and for the loan of photographs; to Barry Rickards for his help in deciphering the geology of the Cumbrian walls and for the insights the geologist brings to understanding the landscape; to Jane Jacobs for giving my initial thoughts on place so much enthusiastic encouragement; to my colleagues at the Faculty of Environmental Studies, York University, and at the Faculty of Architecture and Landscape Architecture, University of Toronto; and, as always, to my professional colleagues at Hough, Stansbury + Woodland.

I would like to express my sincere thanks and appreciation to my friend and fellow landscape architect Margaret Kwan for dedicating many hours to the making of the beautiful drawings that fill the book's early chapters. Most of all I want to thank my wife, Bridget, who painstakingly proofread the galleys and made many other contributions to its development too numerous to mention here.

The illustrations were drawn by Margaret Kwan. Except where noted, photographs were taken by the author.

Introduction

Many years ago when I was a student of architecture at the Edinburgh College of Art, I was enjoined by my elders and betters to study the eighteenth-century terraced houses of Edinburgh New Town. This, I was assured by those who knew, was what great architecture was all about. Anxious to learn, I plodded the streets of this neighborhood, braving the cold, the winds, and the mists of Edinburgh's dismal winter weather, studying the proportions of the facades, the gray elegance of their repetitive rhythms, the detailing of windows, floors, pediments, railings, and ashlar masonry. This must indeed be, I thought, the epitome of architectural order and composition. Then one day, responding to a subconscious impulse, I ducked into a side alley and found myself in one of the internal courtyards behind the elegant but deserted street facades. Here was architectural chaos: windows arranged without apparent order or logic, iron fire escapes cascading down building walls, an extraordinary assortment of exterior plumbing, drying longjohns, pajamas, and bedsheets draped over clotheslines that criss-crossed the courtyard, giving the colorful and festive air of bunting. The ultimate heretical thought suddenly loomed into my consciousness. Why was this so much more interesting and alive than the facades I had been sent to admire?

Many years later my search for answers to this question has left me still puzzling. Why are the abandoned but naturally regenerating places one finds everywhere behind the formal civic landscapes of the city so much more interesting than the ones designers are taught to admire and create? What is it about historic cities and old vernacular landscapes that attracts people? Why do modern cities and suburbs and industrial farming landscapes all tend to look the same? And why is so much intellectual effort spent on hiding modern objects like electrical transmission towers in agricultural landscapes while we immortalize the old windmills of a bygone era? So much of our contemporary landscape, in fact, seems to be unattractive and out of step with established views of what is beautiful. I have for years been intrigued by these questions because the issues that arise from them have often not been adequately addressed by the design professions whose motivations stem more from formal design theory rather than from an understanding of how environments work in real life.

The noted American geographer and writer J. B. Jackson, has described the "sense of place" as one of the ways in which we identify the peculiar characteristics of a landscape and its inhabitants.[1] This book is about what makes the sense of place; how the forces of human and nonhuman nature have, in the past, created characteristic and distinctly identifiable landscapes and how they are shaping the postindustrial landscape today. It is a search for an environmental view of design that recognizes the realities of the contemporary scene, that draws its inspiration from the ecological and cultural lessons of the vernacular, and that emphasizes the need for a sustainable view of the future. It is based on the conviction that, in the context of contemporary life, the sense of identity and place is unique and significant in the shaping of the human environment.

Few people would quarrel with the notion that natural landscapes are the result of biophysical processes that shape the land and create the unmistakable differences between one place and another. Whether mountains, prairies, sea coasts, badlands, or deserts, they are expressions of the region. Similarly, human landscapes and settlements are the consequence of culture modifying and imposing its needs on natural or wild places. Yet the spread of cities since the industrial revolution seems increasingly to have negated these differences. One modern city tends to be very similar to another. The influences that at one time gave uniqueness to place— the response of built form to climate, local building materials, and craftsmanship, for instance—are today becoming obscured as technology makes materials universally available and as climate is controlled by artificially modifying the interior environment of buildings. Technology

has increasingly overcome the limitations imposed by nature. The question of regional character has become a question of choice and, therefore, of design rather than of necessity.

The development of a design philosophy that recognizes diversity and the differences between places is, as Odum suggests, central to the maintenance and enhancement of social and environmental health, since diversity and health are linked.[2] Yet, while traditional vernacular landscapes usually represent the diverse character of different places, conscious planning and design tend to negate those differences.

Attitudes and perceptions of the environment expressed in town planning have been more concerned with utopian ideals and design doctrine than with natural process as a basis for form. Predetermined visions of how cities and human institutions should work have had enormous influence on the contemporary landscape. Yet it can be argued that utopian ideals work directly against natural and cultural diversity.

The perceptual distinction between what is urban and what is rural is a nostalgic view held by city dwellers who, seeking rural quiet or cheap real estate, are themselves the cause of its disappearance. Urbanism is a fact of life in postindustrial society. The idea that one can live a rural life, in the sense of having a working environment distinct from the city's influence, is no longer a valid concept in the developed world. For the farmer holding down two jobs in the urban fringe, or the Inuit hunting in the far north, the realities of urbanism are undeniable.[3] The pervasive influence of the city in every corner of our living environment is challenging our conventional perceptions of the sense of region, of belonging to a place.

Coupled with the evolution of old landscapes into new is a nostalgia for the ones that are being lost, especially in North America where the preservation of historic places has assumed high priority. But in the absence of contemporary relevance or historical continuity these landscapes tell us little of past events or how people lived. Yet it is exactly this nostalgic search for natural, historic, and cultural patterns in the landscape that appeals to visitors. Looking for what is distinctive and different is what tourism is all about. Rocky Mountain scenery, rural settlements, subtropical coastlines and vegetation: these all have irresistible attractions as places to visit. Few tourists purposefully visit the suburbs and industrial areas of large cities. Tourism has become the very life-blood of picturesque but resource-poor or economically destitute countries. Paradoxically, it helps destroy the very nature of the places and social customs on which it thrives.

Modern society presents further challenges to regional identity. Cultural transfer in multicultural societies such as Canada, the United States, and increasingly Europe enhances the diversity and character of many cities. Ethnic neighborhoods are admired for their character, color, and cultural differences. Yet the traditions and styles of living that make them different are constantly threatened by pressure to conform to establish majority values.

HOW CAN insights derived from natural and cultural processes provide us with ways of re-establishing the identity and uniqueness of places in the contemporary landscape? This book explores this and related questions. Chapter 1 identifies the key ingredients of regional identity. Chapters 2 and 3 deal with the natural and cultural processes that form the basis for regional identity; the dimensions of the regional imperative are explored here and conclusions are drawn about its essential nature. Chapter 4 deals with utopias: how values and the way landscapes are understood have influenced attitudes toward the environment. Chapters 5, 6, and 7 explore contemporary issues and the patterns of change that are shaping our landscape today. Chapter 8 deals with design principles and states the philosophy of the book: how the design process can create healthy and clearly identifiable contemporary environments.

1 / The Regional Imperative

Understanding places begins with feelings. Names conjure up a kaleidoscope of distinct or unique sensory images; random, disconnected smells, sounds, and sights crowd our memories. Yet feelings about places differ, depending on whether one is visiting, working, or living in them. The places I remember best are cities like Istanbul where I grew up and which I revisited some fifty years later; Edinburgh where I studied architecture; Hong Kong where I worked as a consultant; and Toronto where I have lived with my family for many years. Memories of places such as these are important to all of us, so it is here that my search to understand the nature of the regional imperative must begin.

Images of Places Remembered

My early recollections of Istanbul are scattered but vivid; the hillside terraced garden of the house where my sister and I used to play; the big fig tree and the swing that hung from one of its branches; the distant view of the Bosphorus where we could see ships at anchor; the shapes of mosque and minaret silhouetted against the background of sea and sky in the morning haze. I remember too the bustling harbor by the Galata Bridge, where we took the ferry to spend our summer holidays on the coast; the rowboats for hire that took

us across the Golden Horn for a few lire; the strange bulbous oars that the muscled and sweating oarsman grasped as he rowed. There were the fishermen on the quay, lines disappearing into murky water; the shoeshine boys squatting behind ornate polished brass stands with brushes, each with a brass handle, gleaming in neat rows. There were the countless cats that roamed the city streets. I sometimes saw them being picked up and taken away by men with large bags; but nobody could tell me just where the cats were taken. Then there were the smells of the sea; the aroma of food cooking and of sesame seed bread being sold on street corners by small boys; the pungent and overpowering smell of feet in the mosques we visited. And there were sounds of the street sellers, the constant roar of traffic, the slap, slap of water against the sides of boats and quays, the sonorous and penetrating horns of departing ferries echoing across the water; the strange, haunting chant of the muezzins calling the faithful to prayer from distant minarets. And I used to wonder how they managed to climb all those stairs to the top, five times a day.

I visited Istanbul again as an adult, eager to see what I could remember of childhood days, my antennae trained to grasp the sensations that had made this ancient city so memorable. This time, however, I saw it as an observer with a knowledge of its history and geography giving my impressions shape and focus. I knew that Istanbul stands on seven hills at the junction of the Bosphorus and the Sea of Marmara, and that the natural harbor of the Golden Horn is unaffected by tides and deep enough for large ships to anchor. Located at the intersection of two great highways of commerce between the Black Sea and the Mediterranean it has been for thousands of years the meeting place between East and West. Historically it was the eastern capital of the Roman Empire and was later founded as the center of eastern Christianity by the emperor Constantine in A. D. 323. As the birthplace of Byzantine art and architecture it influenced all parts of the Eastern Roman Empire: Greece, Russia, Asia Minor, North Africa and, farther west, Venice and Ravenna.

The city has a powerful physical and cultural identity. From the airport highway that flanks the Bosphorus one can see ships from all nations at anchor waiting to unload and take on cargo—a clear indication of the city's continuing importance as a major port and of the significance of its location. The Roman aqueduct through which the highway passes and the old Roman city walls casually avoided bear witness to a modern city building on the shoulders of its past, old and new existing unselfconsciously side by side. The hills on which Istanbul is built, the harbor on which it focuses, and the great monu-

ments that reflect its history are compelling images of the city's presence. From the hilltop site of the Topkapi Palace, built by the Turkish sultan Mehmet the Second in 1473, the panorama of rooftops, sea, ships, and sky stretching out to the horizon expresses the dramatic connection between the city and its landscape.

Istanbul's historic monuments and its dramatic site are only two aspects of its identity: another equally important one is the everyday life of ordinary people going about their daily business in the back streets, squares, and markets that continue to evolve organically through time and circumstance. And it is by the Galata Bridge, which links the city across the Golden Horn, that the real life, identity, and drama of Istanbul are felt. It was there amidst the aromas that filled the air that my early memories came surging back. This is the hub of the old city, a place teeming with life. People, traffic, and boats become a collage of intense energy and activity, of color, sounds, and smells. Cars, buses, taxis, donkeys, and handcarts jam the streets; an endless throng of people cross the bridge and fill pedestrian spaces. The shoeshine boys still squat behind their brass stands and brushes; the fishermen are still fishing in murky waters that look as if they could not even support red worms; the cats still roam streets and markets earning a living from the wastes of the city, and the smell of feet in the mosques seems stronger

than ever. Ferry boats at the dockside loom large over the traffic-filled streets, six abreast as they take on and disgorge passengers. And over this scene the great mosques, historic symbols of religious power and Byzantine culture, still dominate the skyline, the characteristic silhouettes of dome and minaret reshaping the form of the natural hillside. But today the minarets broadcast the call to prayer from tape recorders and loudspeakers mounted on the balconies where the muezzin once stood.

Street life in Istanbul shapes the place. Buildings and spaces are mere stage settings for social and business life. Impressions are legion: the street as commercial and social space; wares of every kind displayed on sidewalks and monopolizing streets, the sheer dominance of people keeping cars out. Store merchandise, clothes, chairs, baskets, copper, and carpets create the complex architecture of the street environment. The aroma of specialities like baklava, turkish delight, and shish kabob delight the nose and tempt the palate. Canvas awnings are strung between opposing buildings to protect the life on the street below from the sun's

A view of Istanbul from the Topkapi palace showing dramatic connections between city and landscape.

Istanbul.
The Galata bridge
and the hub of the
old city, a place
teeming with life.

Istanbul.
Another view
of the Galata
bridge and the
great mosques,
their presence
dominating the
city's skyline and
symbolizing its ex-
traordinary magic
as a place.

The street life and markets in Istanbul shape the place.

Turkish shoeshine boys ply their trade on the city's sidewalks.

fierce heat. The Turkish character is nowhere better expressed than in the taxi driver, highly competitive but friendly and a marked contrast to the unadulterated bad temper of the Paris taxi driver or the hostility of the New York cabby. The complexity of life comes from the interaction of countless individual human purposes which shape and transcend the relative monotony of the surrounding built environment. In Istanbul, the interaction of landscape, a long cultural and commercial history, and a sense of dynamic continuity give the city its unforgettable drama and mystery.

HONG KONG, founded by the British in 1841, is another city of extraordinary presence and sense of regional identity. Like Istanbul, its port is strategically located on the main trade routes between Europe and the Far East, yet its character remains fundamentally different. Its deep-water harbor, the mountainous terrain of Hong Kong island, and the flatter landscape of Kowloon across the bay make up the city's setting, which is best revealed from the top of Victoria Peak. From there one sees a dramatic panorama of green mountains, high-rise buildings clustered tightly on the island's narrow coastal plain, the great harbor sheltering ships from all nations and of every size and shape, and beyond, the city on the Kowloon mainland.

Hong Kong has few historic buildings to identify it with its past. It is a city preoccupied with the present. With six million people, primarily Chinese, housed on just over one thousand square kilometers, Hong Kong has one of the highest urban densities in the world. From ground level, one's first glimpse is of a forest of thirty-story apartment buildings, packed so closely together that tenants could almost shake hands from neighboring apartments. Against the precipitous mountains rising through mists and shifting clouds stand the bland, modern, high-rise architecture of the city center's banks and corporations and the block-long, multilevel shopping concourses. The site, powerful and awe inspiring, dominates the city yet remains a backdrop, disembodied from an urban form that expresses contemporary preoccupations with finance and trade, not with setting. From a distance, Hong Kong's urban character could be anywhere.

The essence of Hong Kong, however, can be found in the web of streets that lie at the bottom of the skyscraper canyons and beyond the Western-style shopping plazas. The life of the city is here, a world of constant buying and selling, of camera shops, jewelry stores, fruit and vegetable markets, of endless food stalls where a meal can be had for a few cents. It is a place where men with umbrellas for sale suddenly ap-

Hong Kong from Victoria Peak. The dramatic interaction of city, sea, and landscape creates its unmistakable identity.

Hong Kong. Viewed from the harbor, the backdrop of the mountains dominates the city.

Hong Kong's working waterfront. Chinese junks and other craft ply the coastal waters carrying goods of all kinds. The working barges in the background service the port.

Hong Kong. Thousands of Chinese spend their lives afloat on every conceivable kind of craft. For the visitor the scene has an intensity, variety, and color that come from the necessity of making do in a desperately overcrowded city.

pear the moment it begins to rain; where advertising signs form a canopy over streets, creating a gaudy and vibrant nighttime world of color and astonishing calligraphic artistry; where streets are so packed with people that cars only enter on sufferance. Looking over the city from the upper floors of offices and hotels one sees ramshackle huts commandeering the rooftops of newer buildings, the skyline a forest of television antennae. Squatter's shacks cascade down the mountain slopes, a poignant reminder of Hong Kong's housing predicament. There are the apartment forests in the newer parts of the city, their balconies and facades festooned with potted plants, bric-a-brac and the pageantry of washing hanging out to dry. In the

harbor, behind the breakwaters built to protect boats from typhoons, another environment unfolds—that of life on the water: junks, houseboats, rafts, and almost anything else that can be lived in, are crowded together in a mass of floating humanity that is breathtaking in its density, complexity, and color.

Victoria Park, the one major open space where the city is kept at bay, is thronged with people from daybreak to midnight. The young jog, play football, tennis, and basketball; older men congregate to compare the relative merits of caged songbirds; a gaggle of elderly Chinese ladies perch on the children's swings and seesaws, gossiping; platoons of people of all ages practice Tai Chi in collective slow motion. There is a sharing of

Life and commerce in Hong Kong's streets: constant buying and selling and the necessities of survival are a way of life.

Leisure in Hong Kong's only large public space: Tai Chi at 6 A.M. in Victoria Park.

space with a sense of personal privacy. The park, lushly green at canopy level, has the marks of a place much loved and relentlessly used, with different places for different activities—strolling paths, flower displays, places for active sports, nooks and crannies for meditation.

Hong Kong's regional identity, then, exists in two extremes: in the drama of its magnificent landscape of coastline and mountain and in the everyday goings-on of the city streets and urban spaces that express collective aspirations and the struggle to make do in the most crowded city in the world.

Canada was once described by the London magazine *The Economist* as the second dullest country in the world in terms of "the yawn factor."[1] Notwithstanding the inaccuracy of such a charge today, Toronto was at one time similarly branded by visitors to the city. It was not an exciting place to visit. People who live and work there, however, have a very different point of view. What is boring for the visitor is quite the opposite for the resident. Toronto's sense of place is not derived from a dramatic landscape or from a distinguished historic architectural heritage. It has neither. Unlike modern Hong Kong, its essentially flat landscape denies any hint of its location on Lake Ontario, and there are no views at ground level of the lake except at the water's edge. Compared with historic Istanbul, its urban fabric is un-

distinguished, the architectural styles of downtown reflecting an international preoccupation with finance and business. Its identity lies elsewhere—in the familiarity and sense of security that comes from those who have lived there for a long time; who know its nooks and crannies, the best places to shop for a bargain; the favorite places for meeting people. It is found in the ethnic neighborhoods that have street signs in their own language as well as in English; in the commitment of local communities to protecting their neighborhoods; in the safety of its streets; and in the involvement of people in the everyday affairs of the city.

As a long-term resident I have fought for the preservation of the local ravine where I, along with many others, walk the dog on a Sunday. I have campaigned to protect our local streets from through traffic and worked to stop a proposed expressway that would have devastated the city's fabric. As a long-term resident, I have a commitment to the place. My neighborhood is as ordinary as one will find in any North American city. But the street where I live has become part of my front yard. It is important for the everyday contact among neighbors, for the children who play there together, and for the sense that that's where I belong. The yearly street party is a symbol of our collective investment in the place. It is a sense of identity that is not apparent to the casual visitor who rides through the streets of the city in an air-conditioned tourbus or arrives with five hundred others for a convention. For all its sense of the ordinary, Toronto is a very real and vital place for the people who live there.

These three cities are each memorable and unique in many ways, in location, language, and culture, and in the ways they are perceived. Yet they also have common themes that are important to the way we understand regional identity in other places.

The Elements of Perceived Identity

Landscape is an expression of a place's regional context especially in the absence of distinguishing architectural styles. Wherever one goes in Hong Kong, the grandeur of its setting—mountain, sea, and enclosing harbor—and its tropical climate and vegetation are the most clearly understood expressions of the place. For the most part, though, site does not shape contemporary urban form. Like San Francisco, it is there in spite of the city. By contrast, many older cities have an undeniable sense of identity because site and urban form, architecture and landscape, become one. A key to Istanbul's identity lies in its skyline where topography and architectural form are fused into a single symbolic expression of culture and nature. There is a sense of a city adapting itself to the constraints

The many faces of Toronto.

The Kensington neighborhood is famous for its street market.

Chinatown. Another ethnic neighborhood that expresses the cultural diversity of the city.

The Greek quarter. The city's different places are identified by the street signs, each in the language of the particular neighborhood.

A street party in a residential neighborhood. Places that may seem very ordinary to an outsider have a special identity for the local residents.

and opportunities the landscape affords.

Edinburgh, another city famous for the way the built and natural environments come together, has this characteristic. The city seems to emerge from its landscape. There is the long massive rock dominated by the castle; the ridge along which the old medieval town grew linking the castle with Holyrood Palace. There is Princes Street and the sweep of the valley and its gardens, contained on the north by the wall of buildings and shops of the street facade, and on the south by the brooding presence of the castle. There are the views along Princes Street ending in Calton Hill and its nineteenth-century Greek temple ruin, called "The Folly" by students in my day. The magic of this city is contained by the marvelous interplay of its topography, built form, and open spaces.

Pedigreed architecture—defined by Bernard Rudofsky as that style of building that is based on formal rules of design and that commemorates power and wealth[2]—symbolizes broader aesthetic preoccupations of different cultures and different periods of history, whether these are dedicated to religion, the military, art, urban design, great gardens, or commerce. Pedigree is the outward and formal manifestation of the way a culture perceives its identity. The unique religious architecture of the Byzantine church in Turkey, Africa, and Italy; the essential differences

between English and French cathedrals; the Renaissance gardens of Italy and France are all distinctive and well-known expressions of cultural identity in the context of time and place. Similarly, many beautiful towns, like Siena and Bath, have combined a superb use of their sites with gracious architecture and monuments, while others, such as Florence, have maintained a distinct identity without dramatic natural features. The great monuments that punctuate city skylines—St. Paul's in London, or the Blue Mosque in Istanbul—or rural skylines—Salisbury Cathedral—are the major features distinguishing one place from another.

A view of the "new town" from the castle heights. (British Tourist Authority)

Edinburgh.

A key to Edinburgh's identity lies in the fusion of skyline, topography, and architecture into a symbolic expression of culture and nature. Here, a dramatic view of the city looking west. The castle rock and Princes Street frame the valley and Princes Street Gardens, with the spire of the Walter Scott monument creating a dramatic focal point. (British Tourist Authority)

A view of the castle overlooking Princes Street Gardens. (British Tourist Authority)

Another characteristic of the pedigree is the use of materials that have no connection with the local site. Marbles from Italy, hardwoods brought from the Far East, or exotic plants imported from alien climates have always been an expression of wealth and influence and a denial of regionalism as defined by the vernacular. Denial of regional identity is, in fact, not restricted to modern times but is inherent in most pedigreed forms. There are, for instance, few places less expressive of the native French landscape than the palace and gardens of Versailles.

Beneath the pedigree of architectural styles and fashions are the urban forms that develop in response to the practical necessities of daily life. The vernacular contains a greater sense of the place than the veneer of wealth and power that have long been the subject of history books and holiday tours. The simplicity and directness of form of old places that convey a sense of having grown out of the landscape are, to a great extent, the consequence of how local materials were used and of the need to solve problems of shelter and security as expeditiously as possible.

In this context, regional identity is a notion of something that is experienced. Some of the images that I have discussed are from places I have lived in, others are from places I have visited. How one perceives the identity of a region will depend on whether one is a resident or a visitor, and for each there are different ways of understanding it. A place may not have a spectacular landscape, or great architecture, or even a long history. The markers, reference points, boundaries, and other symbols of everyday experience may be unseen or not understood by outsiders. But for those who live there these things are what give a place its meaning and relevance.

The names people give to places imbue them with a symbolic significance that unnamed places lack. In Toronto, street signs in the Greek quarter or in Chinatown give a stamp to these neighborhoods that defines and reinforces their identity within the city. Naming is, in fact, endemic to the perceptions and shaping of a locality, for names alone create a mental image that has special significance for local people—and names can be the means by which an outsider begins to perceive a place's unique qualities. Grady Clay describes place as "a creation, a byproduct, a resultant of human occupancy and presence . . . a cliff is not a 'Place' until it becomes a destination, like *Lover's Leap*."[3] When I arrived in Edinburgh as a student I spent my first day walking and exploring its many monuments, streets, and squares. On Princes Street I stopped an elderly Scotsman. "What is the name of that castle on the hill?" I asked politely. The old man stared at me. "It's no' *that* castle," he replied, clearly incredulous at my ignorance;

"It's *The Castle.*" I went away chastened by my stupidity, but seeing Edinburgh in a new light.

Names may also create strong and romantic images of places that belie reality. Piccadilly Circus is one of the world's great places people go to see when they visit London. Until recently, though, the famous statue of Eros was almost totally inaccessible because the deafening traffic that circled the place made it a death trap for pedestrians. As a friend of mine once bluntly described it, "The name's magic, but the place is the pits." In whatever form they occur, however, the names of places, and for that matter most things we value, are the essential and powerful features of their identity and how they are understood.

THE ISSUES surrounding the regional imperative have immense environmental, cultural, and social significance. Ancient cities that have evolved over hundreds of years are immediately recognizable in ways that modern cities are not. Yet from the way that even these old cities have expanded over the past thirty to forty years, it is abundantly clear that lack of identity has become a universal phenomenon. It pervades not only cities everywhere but also the entire regional landscape. The market forces of trade, economics, and profit, and the technologies to master uncertainty and create security of supply are the imperatives driving contemporary cities.

A basic principle of habitat building in both human and nonhuman communities is that where the processes of nature are the same, similar forms will emerge. As Bernard Rudofski has observed, "Altogether, cities correspond closely to the ideas and ideals of their inhabitants. [Modern cities] are the tangible expression of a nation's spirit or lack of spirit. . . . In Lyndon Johnson's cheerful view they are 'the result of greed and stupidity.'"[4] The principle of common greed, however, is probably no different today than it ever was. Amsterdam was built by business interests, and as Mumford points out, "[the city] bears witness to the commercial spirit at its best. . . . The fact that it was not widely imitated shows that it was not capitalism alone, but complex institutions, personalities, and opportunities, coming together at a unique moment, that made the city one of the greatest examples of the town planner's art. Even so, it remains capitalism's one outstanding urban achievement, rivaled only by elegant Bath."[5] But today capitalism and economics seem to be the primary sources of inspiration for urban form, unenriched by past traditions of urban design. Irrespective of the civilizing influences of cultural tradition, however, it is the native landscape that is a primary determinant of regional identity.

Natural scenery has a powerful influence on our perception of places, and we experience it through all the senses. The differences among the mountains, deserts, prairies, and forest regions across the North American continent, for example, are both obvious and compelling to any traveler. Similarly, it is plain that the scenery of coastal landscapes differs in Maine, California, southern England, and the Bahamas. There are traits in these environments to which we respond emotionally, in ways that are affected by our culture and background, and by ephemeral events such as the patterns of sunshine and cloud, mist and rain, or the colors of the seasons.

Nature sets a diverse stage for the aesthetic inherent in every landscape. Whether it is the silence and rugged grandeur of snowcapped mountains in the Rockies, or the great sweep of ocean waves stretching into the remote distance along the Oregon coastline, or the rushing torrent of an upland stream in the Cape Breton Highlands of Nova Scotia, the drama of natural scenery strikes a primeval chord in all of us. Consciously, or unconsciously, there is a sense of wonder at what confronts us, a feeling that we are face to face with overpowering natural forces, of being overwhelmed by the presence of the place.

How landscapes are appreciated

The Oregon coastline, where the sea meets the land. Natural processes create powerful and memorable images of wild and beautiful landscapes, as well as an immense variety of form and character.

Georgian Bay, Ontario.
A Precambrian shield landscape of rocky islands, wind-blown pine, and blue water that has become a symbol of the Canadian landscape inspiring poets and painters.

Banff, Alberta. The grandeur of the Rocky Mountains is expressed in their coniferous forests, snow-capped peaks, and cascading rivers.

Several influences act on our responses to the imagery of natural landscapes.

Spectacular scenery that combines the contrasts of topography, water, and forests is exciting because of its obvious variety and drama as rugged terrain. It is easy to admire.

Representative landscapes are often overlooked, either because they lack this obvious variety and drama, or because they are so familiar.

A natural feature is easier to appreciate when it can be seen in the context of its surroundings where contrasts are apparent.

A walk through a varied landscape maintains interest when the spatial experience, natural features, and detail occur in quick succession.

The sheer scale of a uniform landscape continuing uninterrupted over large distances implies monotony. The scale and appreciable detail of landscapes are related to the type and speed of travel, which determine how we perceive the world around us.

and what is enjoyed aesthetically have been subject to radical transformation over the centuries. Our perception of scenery has been greatly influenced by poets and painters. Poussin's grandiose mountain scenes taught eighteenth-century society to see landscapes in terms of the picturesque. The lakes and mountains of Switzerland and northern Italy, previously endured by the necessity of travel, now became something to be admired[1] so that a hundred years ago, Ruskin would proclaim, "Mountains are the beginning and end of all natural scenery."[2]

George Seddon has remarked how the Sichuan Alps on the China-Tibet border show an uncanny likeness to traditional Chinese landscape paintings—a case of nature imitating art.[3] Chiang Yee's artistic interpretations of the English Lake District[4] are founded on his inner vision of familiar Chinese landscapes that are totally different from a Western painter's view of similar scenery—such as one sees in Turner's paintings.

Since the Second World War, the increasingly fashionable paintings of the Canadian Group of Seven have made the public aware of the beauty, drama, and wildness of Ontario's Georgian Bay landscape and so created an appreciation for wild scenery that had not previously existed. And as the private car has made travel and summer cottaging available to a larger group of people, awareness and appreciation of the region has increased. People have been taught to see the scenic landscape through the eyes of the artist.

Nicholas Poussin. Landscape with a Woman Washing Her Feet *(National Gallery of Canada, Ottawa. Gift of H. S. Southam, Ottawa 1944). Poussin's paintings of the Italian countryside, describing nature's various moods, brought a new perception of the romantic landscape to eighteenth-century society. They greatly influenced the picturesque garden movement.*

Chiang Yee.
The Silent Traveller in Lakeland. *The Chinese artist's paintings are more reminiscent of the processes that shaped his native mountains than of those of the English Lake district.*

Originally thought by Westerners to be highly stylized artistic interpretations of Asian mountains, Chinese paintings have been found to represent accurately the geomorphology of these landscapes.

Perceptions of what is beautiful, however, have tended to focus on the spectacular and the dramatic—on scenery that is popularized in the picture postcard and coffee-table picture book. A holiday experienced through the windshield of a car, or from the window of an air-conditioned bus is later remembered in the mandatory slide show at home. There is very little room for ordinary, representative places that require another kind of appreciation. The preoccupation with scenery as visual enjoyment is an expression of society's dissociation from nature and the processes that shape the land and its scenic variety. As Yi-Fu Tuan

notes, "[Scenery] can seldom command much emotion. The scenic spot along the highway presents us with a picture window of nature which, sublime as it often is, cannot move us to any response more strenuous than the taking of a snapshot."[5]

The need to measure and quantify the aesthetic superiority of one landscape over another that became a required tool for resource planning in the 1970s demonstrates one aspect of the problem. Objective criteria for selecting superior sites are necessary to convince decision makers that a particular landscape is worthy of preservation. Scenic planning in-

Tom Thomson. The Jack Pine (National Gallery of Canada, Ottawa). Through their paintings the Canadian Group of Seven influenced the way people perceive and appreciate wild scenery.

volves creating systems for measuring the permanent elements of a landscape. But these systems necessarily leave out the unmeasurable, ephemeral things that in reality are largely responsible for the aesthetic experience. They also leave out environments such as prairie landscapes that can be perceived as having little or no aesthetic value, and which, therefore, could be crossed off the list of significant places. As Neil Evernden has observed, "Low ratings may accurately indicate the public's low opinion of the prairie, but the ratings also point to a central feature of this landscape, the absence of things. It is a feature that seems to condemn this landscape to aesthetic poverty. Nothing is there to measure and enjoy. There is nothing to possess aesthetic interest, so how could the scene be beautiful?"[6]

The classification of scenic landscapes and our perceptions of their beauty or lack of it are based on the notion that such places have only recreational uses. The trip to the seashore is made by people who don't live there. A comment by the conductor to the passengers who were staring out of the window as the cross-Canada train rattled through the prairies that "there's nothing to see out there"[7] could easily have been made for the benefit of tourists on their way to the Rockies. It is arguable that there isn't much to see from the train window. But for the people returning home to the prairies the perception would be quite different. For them it's home, a part of everyday life and familiar territory, a place where time and emotion have been invested. For these people the conductor's statement is meaningless. Understood close up, the prairies are full of varied detail: the changes in weather bring shifting ephemeral images of the landscape; the sky dominates the flat plain. There are all kinds of things to see and know there. It is a question of how they are experienced.

Natural Process: A Foundation for the Regional Imperative

It is clear that the basis for aesthetic experience and understanding of natural landscapes must rest on a firmer footing than quantified measurement or mere fashion. Emotional appreciation as a subjective reaction to beautiful and varied scenery must be understood through the natural processes that give rise to it. Appreciation of scenery increases the more we delve into natural history, which tells us about things we cannot see as well as those we can. Our apprecia-

tion of any art form, whether music, painting, poetry, or ballet, increases in proportion to our knowledge of its underlying structure. This is also true of natural process and its visible manifestations in the landscape. The closer we look, the more compelling the beauty and variety. Thus, for the general public, bringing aesthetic appreciation and natural history together is important for an understanding of places and for the development of a new and non-consumptive attitude to the regional landscape. For the designer concerned with the land, such a holistic view is essential.

The shape of the land is the product of natural history—of the forces of mountain building and continental drift; of climate influenced and modified by geography and topographic features; of erosion from water, wind, and glaciers. The living world of plants and animals flourishes in an infinite number of places and in an infinite variety of shapes, behavior, and adaptations. Each has its own inherent character, which is distinguishable from one place to the next. Regional differences may be both more and less obvious. While the differences in a typical mixed hardwood forest may not, at first glance, be readily observable, a closer look allows the variety in the detail of its flora, rocks, and streams to emerge. Two places that I have known in different parts of the world illustrate the legacy of natural processes in shaping regional variety, as well as the need to understand the aesthetic of places in the context of their natural history. Both are coastal landscapes: one is small and discrete at an easily understood scale; the other is immense and continuous at a scale not easily comprehended.

The French island of Guadeloupe in the Lesser Antilles is a remarkable place for the natural diversity of its tiny island mass. Set in the tropical zone of the Caribbean and tempered throughout the year by the tradewinds, it is sought after by tourists for its sunshine, balmy climate, and beautiful beaches, as well as its local and French cooking. There are, in fact, two principal islands, separated by only a narrow strait: Grande Terre to the east and Basse Terre to the west. The two islands differ from each other radically, in geology, land form, vegetation, and overall character. Grande Terre is arid with little topographic variation. There are virtually no natural rivers or streams on the island. Apart from the coconut groves that line the beaches, its vegetation is sparse and little remains of the original natural cover. Most of this has been cleared for sugar plantations over successive periods of British and French occupation. The underlying soil structure is calcareous, which accounts for the fact that the island's beaches are made of pure white sand and are protected from the ocean waves by coral reefs of exquisite beauty and diversity.

Guadeloupe. Different places have been created by different natural processes. White calcareous sand beaches, coconut groves, and flat terrain occur on Grande Terre's coastline.

Guadeloupe. In contrast, black volcanic sand beaches and mountainous terrain are characteristic of Basse Terre's coasts.

The total picture of white sand, coconut palms, and intense deep blue sea is typical of the magical image of the Caribbean islands. The entire focus of Grande Terre is toward its coastline, a fact that is clearly evident in its historic dependence on the sea for food and defense and, today, for tourism.

Just fifteen minutes drive from the capital, Pointe-a-Pitre, is the other island, Basse Terre, a steep mountainous terrain created by volcanic action that continues today in the still-active volcano La Soufrière. In contrast to Grande Terre, the coastal beaches are composed of ground-up

In Basse Terre's interior is another landscape of spectacular falls, cascading rivers, and tropical vegetation.

rainfall, and community composition.

The land use of the two adjoining islands of Guadeloupe, so dramatically different in geomorphology, soils, and plant life, reflects their natural diversity and defined character. Grand Terre's coastline is the focus for settlement and tourism, while much of Basse Terre has been designated as natural park, devoted to conservation and education. In every respect the two islands stand in marked contrast to the other coastal landscape in this comparison—the Hudson Bay lowlands of northern Canada.

THE WEST coast of Hudson Bay is a vast flat region of tundra, water, boulders, and shallow rivers that tell of its recent glacial origins. The ever-present dome of the sky is reflected in a watery landscape that is both powerful and vast in scale. It is a land of permafrost where the thin soil thaws to a depth of a few centimeters during the short summer when the land bursts into violent life.

This landscape is the consequence of the ice sheets that advanced and receded over two million years ago. The sheer weight of thousands of meters of glacial ice had the effect of depressing the earth's crust, which, following the retreat of the last glacier some twenty thousand to thirty thousand years ago, has begun to rebound. Thus the bay is slowly giving way to land, and it is estimated that at some time in the future, as-

volcanic rock and are totally black in color, giving an entirely different character to this island's shoreline. The mountain rises to a height of 1,400 meters above sea level. It is a landscape of steep lush valleys, cascading rivers and waterfalls that drop hundreds of meters, and verdant tropical forests. These forests change dramatically as each change in level, from summit to coast, creates distinct geographic divisions in temperature,

suming that another ice age does not occur, it may disappear altogether. The receding glaciers left shallow ponds and lakes, large granite boulders and eskers that resemble a continuous line of rock, sand, and gravel thrown into a heap by a giant and stretching for miles across the tundra. Even in its friendly moods this vast, watery landscape feels slightly menacing. Its most dominant aspect is the great, uninterrupted dome of the sky. Standing in this landscape kilometers from camp, I had the sense of its vastness and power and of the insignificance of man. From an elevated feature such as a boulder or an esker, however, the sweeping views that these formations provide create a definite psychological advantage and make the landscape more understandable. The sounds of civilization are totally absent in this wilderness. Only those of nature are heard: the loons at night, other birds by day; the constant sound of running water over rocks; the wind; the buzz of the ever present mosquitoes.

As one looks over this flat landscape, it becomes clear that it is composed of very different environments. The tundra is flat with shallow ponds and lakes predominating. The ponds contain sedges, buttercup, potentilla, and other aquatic plants growing in a combination of soft mud and soggy vegetation that has hardly begun to decompose. On slightly higher ground where sand outwashes have produced better drained soils, a variety of quite different plants flourish. Mosses, grasses, and brilliant splashes of color greet the eye— white ledum, mountain avens, and andromeda. Birch and willow, although they are trees, lie close to the ground in this harsh environment, in response to nutrient-poor soils, a short growing season, and the ever present wind that sweeps unchecked across this unprotected landscape.

Wildlife abounds and animal adaptation to the environment is direct. The caribou seek out the coast breezes for relief from the hordes of ravenous mosquitoes. A lone arctic wolf investigates the camp from a safe distance before trotting away. The ground squirrels sit down to breakfast with us at our camp every morning and enjoy the unaccustomed luxury of left-over porridge and anything else they can take without being noticed. These animals live in colonies on the sandy ridges and eskers, burrowing just below the sod roots where digging is easy and maximum insulation from the cold is assured. White arctic daisies colonize the disturbed soil around their burrows—a clear indication from a distance of the presence of a colony. Adaptation to the environment in the interests of survival is very apparent in these direct responses to soil, topography, wind, and availability of food.

This northern landscape may appear bleak, endless, unremarkable and, to all intents and purposes, de-

The west coast of Hudson Bay. Minor variations in land form create major differences in vegetation, habitat, and character. The drama and beauty of this landscape lie in its detail.

*Hudson Bay. One of the louseworts (*Pedicularis soudatica*) growing in marshy sites.*

A sandy outwash presents a different environment of boulders, and dryer, gravelly soils and plants.

void of variety to people who see it for the first time. But its overall character, its geomorphology, water regime, soils and plants, and animals and birds express a regional personality every bit as powerful as the Guadeloupe landscape. Its drama and beauty, however, lie primarily in its detail, in the complex variety and interdependence that become apparent when one begins to understand the natural forces that shape it. Regional differences in landscapes undisturbed by human presence are thus composed of many scales that vary from one climatic region to the next, as well as from place to place. Whether one is looking at the larger regional landscape or at local differences, the forces of nature tend toward complexity of form and species, toward differences between one environment and another, and toward life forms that are uniquely adapted to their location.

Natural Patterns of Similarity

Similar natural forces create generally similar landscapes. Relatively young mountain ranges all have the same general patterns of form whether they are the Rockies, the Alps, or the Himalayas. Valleys formed by glacial action in mountainous regions all have a similar U-shaped form. Eskers and moraines, created by the retreating glaciers, are the same in all northern landscapes. There are correlations between plant communities on a horizontal plain from the equator to the poles and those that occur vertically from sea level to mountain tops. Altitude causes similar differences in natural regions to horizontal distance. As mountains gain altitude the changes in environment at different levels are reflected in changes in plant and animal communities. For instance, mountains in equatorial areas of the world have a microclimate that ranges from jungle heat to glacial cold in a few kilometers. In the Ruwenzori in central Africa, the plant and animal communities in each zone change from savannah to tropical rain forest, to bamboo forest, cold wet heath, and alpine regions where pioneer plants cling to wind-swept rock and glacial edges.[8]

A good example of how dramatic and localized vertical environments can be is Grande Terre on the island of Guadeloupe, discussed earlier. Within a two-to-three-hour walk from the top of the volcano to the coast a number of distinct major ecosystems can be readily experienced. Among them are the high altitude savannahs above 700 meters, the stunted forests drenched by humidity and low temperatures. Below is the dense lush tropical vegetation of the rain forest. Farther down still come the dry zones of thin forest where rainfall is light, sunshine intense, and humidity slight. And at sea level there are the coastal beaches, swamp, and mangrove forest.

Livingston has observed that plains environments are much the same the world over.[9] What makes them different, however, are the species of fauna that inhabit each. The former grasslands of Canada and the savannahs of Africa have a great deal in common besides their appearance. African wildebeests crop the grass in the same way as the North American bison. Antelopes in the African plains share similar characteristics with the unrelated Canadian pronghorn. The prairie meadowlark, which is not represented outside the Americas, is similar in size, coloration, song, and in the way it builds its nest to the African longclaw that comes from quite different origins.[10]

In this chapter I have examined the nonhuman landscape as a basis for regional identity. Few environments, if any, however, can be viewed in isolation from humankind. The necessity of adapting to and modifying the environment is as fundamental to all societies as it is to the nonhuman environment. Even in the Hudson Bay lowlands, a place that superficially seems to belong to "nature," there is evidence of the presence of the Inuit. The circle of stones and litter on the tundra tells of an old campsite, with bits of old clothing, and the remains of a wolf skin, stiff and white from many seasons of exposure. The top of an esker crowned with curious piles of rocks—named "Inuksuk" by its builders, and thought to derive from early hunting practices and to have been used as a point of reference—is an indication of native activities adapting to and influencing the land.

"Nature knows nothing of what we call landscape," said the Englishman William Marshal in the eighteenth century. Nature's scenery is natural habitat, while our landscape is habitat manipulated by man.[11] The perception of the Hudson Bay lowlands as a distinctive environment is enhanced by those who have lived, hunted, and traveled through that region, whose way of life and culture were shaped by its opportunities and constraints. The concept of natural process is an idea that is useful only if one perceives humanity as part of it. Artificial distinctions between what is "human" and what is "natural" that have dogged much environmental thinking in the past become more blurred and less useful as we try to understand the forces that shape the evolving landscape. The aesthetic response to natural phenomena must be seen as a consequence of such processes.

The presence of man: an "Inuk-suk" built on the eskers by the native people of the north.

3 / The Cultural Landscape
Regional Identity by Necessity

People respond to cultural landscapes in the same way that they do to natural scenery. They have little difficulty in recognizing the marked differences between one kind of rural landscape and another, or between old towns in Italy, England, or Greece. The eighteenth-century agricultural patterns of southern England with their enclosed fields, hedgerows and copses, the stone-walled sheep country of the Yorkshire dales, the vineyards and orchards of the Tuscany region of Italy, or the Loire Valley of France are all distinctive rural places that delight the eye with their picturesque beauty, human scale, sense of harmony, and apparent peaceful stability. They have responded to cultural and natural forces that make them very different, spatially, visually, and in feeling. Old towns and cities in their charm, human scale, and fit with the land are also clearly different one from another. In this chapter I explore what lies behind the distinctiveness, sense of place, and beauty of these vernacular landscapes since they offer important lessons in our search for regional identity today.

The Vernacular: Social and Environmental Linkages

Much of what we find interesting and beautiful about the cultural landscape lies in what we call the vernacular. The vernacular has traditionally been described as forms that grow out of the practical needs of the inhabitants of a place and the constraints of site and climate.[1] Vernacular forms are shaped by many forces: the determinants of nature (biophysical processes and climate); the culture and history unique to each place and time; the role of a central authority whose decisions impose an organizational structure on the landscape. The communities that have created them seem to have had few illusions of overall destiny, long-range plans for entire regions, or visions of utopian places. They evolved from necessity, from the need to solve the immediate practical problems of shelter, town building, and making a living from the land. The vernacular is usually (though erroneously) perceived as belonging to the preindustrial era. As Kevin Lynch has observed, the attraction of vernacular places was "usually the consequence of slow development, which occurred within sharp constraints of natural condition and cultural limitation and since then have been enriched by continuous habitation and reformation."[2] Their forms derived from the limitations of agricultural and building technology, native materials, climate, soils, and established traditions.

Culturally, what we see as regional identity has been determined by the social and institutional linkages that tied people to one place and dictated how they should behave and lead their lives. As Ralf Dahrendorf suggests, "Pre-modern societies, with their overpowering forces of family, estate or cast, tribe, church, slavery, or feudal dependence, were, in some ways, all linkage and no choice."[3] The ties that bind a community together are still apparent in many places. As a youth I lived in a small rural village in England where the essential nature of village life was still clearly defined. The local building contractor's family had been in business in the village for generations. His father worked on our house for my father, and his sons carried on the business after him. The same was true for the local farmers. The neighboring farm where we bought our milk had been in the same family for at least three generations. Most of the local people were related to each other in some way. The electrician who wired our house when my wife and I moved back again some years ago was related to the contractor. The old lady up the road from whom we bought our vegetables was related to both of them. And so the intricate web of family connections and village politics maintained a close-knit social landscape that was both understandable and compact. In fact it took my father thirty years to become accepted as part of village society. This is what staying put is all about.

Being tied to the place also means understanding the environment close to where you live but not beyond it. Thomas Hardy writes about this in *Tess of the D'Urbervilles:*

The Vale of Blackmoor was to her the world, and its inhabitants the races thereof. From the gates and stiles of Marlott she had looked down its length in the wondering days of infancy, and what had been a mystery to her then was not much less than mystery to her now. She had seen daily from her chamber-window towers, villages, faint white mansions; above all the town of Shaston standing majestically on its height; its windows shining like lamps in the evening sun. She had hardly ever visited the place, only a small tract even of the Vale and its environs being known to her by close inspection. Much less had she been far outside the valley. Every contour of the surrounding hills was as personal to her as that of her relatives' faces; but for what lay beyond her judgment was dependent on the teachings of the village school.

Being tied to the place involves stability and a sense of investment in the land because one's well being and survival depend on it. It has been said that tribesmen living in the jungles of the Amazon recognize and know the properties of hundreds of plants. Yet to them the concept of botany as a special branch of knowledge has no meaning. It is something more basic, a part of life itself without which life would be impossible.[4]

The preindustrial landscapes were working environments. A symbiotic relationship between land and settlement characterized them. The land produced the food and raw materials for the settlement, which in turn returned the by-products to the land. It was a close-knit physical, social, and economic relationship that existed through necessity and was limited by technology and tradition. The result of this interaction between man and nature was the creation of a distinctive regionalism whose inherent natural character was shaped into a cultural landscape by human activity over generations. The landscape, in turn, drew its distinctive sense of place from the underlying natural patterns of the land. These forces combined to produce landscapes of extraordinary beauty, human scale, and charm, but in our enjoyment of them today, we no longer see or sense the grinding hard work, poverty, and social cost that frequently accompanied their creation.

AMONG THE most interesting characteristics of man's historic modifications to the environment is the way vernacular forms have dramatized the differences inherent in the natural patterns of the land. Cultural and natural history thus combine to create landscape variety. Traveling through the sheep-farming countryside of Cumbria in northern England, for example, one cannot fail to marvel at how visually powerful this landscape can be. It is clear to anyone that this is a unique place.

Greatly influenced by glaciation over the past million years,[5] denuded of forest over hundreds of years, and drenched by rain almost daily, this is a landscape of drama and contrasts. The magnificent rounded and bare shapes of the fells[6] sweep down to broad river valleys where villages and farmsteads huddle together in isolated clusters of habitation. It is a landscape of great scale and grandeur, of distant views and loneliness, of black rain clouds and mists over the rolling hills, of sunlight casting sudden shafts of moving light over field and fell. The obvious distinguishing mark of this landscape is the complex pattern of stone walls, enclosing fields in the valleys and linking stone barns and farmsteads into a continuous and unified pattern of built form.

This countryside is quite different from other places like the gaunt mountains of Scotland. In spite of the similarities of form in the bare hills, moorlands, cascading streams and gorges, the differences lie in the general absence of walls in the Scottish hills. The dry-stone walls superimposed on the Yorkshire dales countryside are the determining cultural features that differentiate two somewhat similar geographic regions created by nature. "Often the only sign of the hand of man [they] symbolize the very soul of the high Pennines, the endless varying patterns of grey against green."[7]

Indeed, the influence of culture on

A view of Sedbergh with the school playing fields and grazing land, and Winder Fell looming in the background. Landscape identity in Cumbria in northern England is created by a combination of dramatic land form, changing patterns of sun, rain, and cloud, and by hundreds of years of sheep farming.

A valley farm at Coldkeld, near Sedbergh, showing early, and small-scale, walled sheep enclosures with the later, and larger, pattern of walls on the upper slopes.

Yosemite Valley, California. Natural mountainous land forms that have been shaped by glacial action are generally similar in character.

The Jungfrau valley in Switzerland. Similar glacial landscapes are made very different when they are overlain by culture.

the creation of unique landscapes from similar natural ones is apparent in many other situations. For instance, the glacial valleys of Switzerland in general form and character resemble the Rocky Mountains and Sierra Nevada mountain formation in North America because they were created by the same geological processes. They are, nevertheless, immediately recognizable as different places through the intervention of culture. Compare the glacial valley of Yosemite National Park and the valley beneath the 4,142 meter mountain of Jungfrau in the Swiss Alps. Yosemite is apparently undisturbed by human presence. The Swiss valley, however, is modified by upland agriculture, farmsteads, and villages.

Conversely, natural forces influence the creation of unique landscapes from similar expressions of culture. For instance, the dry-stone walls that climb the hills of western Turkey and the Yorkshire dales, are similar in their layout and use to define boundaries. Yet the two landscapes have clearly distinct identities. Another example is the contrasting types of environment between the northern and southern regions of England (in terms of their climate, geomorphology, vegetation and soils) that have created distinct regional differences in an otherwise similar sheep-farming economy.

Natural History Revealed through Culture

The walls, barns, and local villages of the vernacular landscape all express a sense of belonging, a directness and timelessness in form and materials; they are evidence of man working on the land, but in ways that are not always immediately apparent. The stone used to build the walled fields of the Yorkshire dales reveals, in addition to the essential character and history of the region, its underlying local geology. If one takes a closer look at their origins, one finds that the walls, farmsteads, and barns that cluster in the valleys or sit isolated and lonely in more remote places are built from the native limestone that underlies the surface of the land and appears in stony outcrops and escarpments in many places on the barren uplands of the dales.

The marvelous harmony, the feeling of belonging and sense of place that is characteristic of this countryside, is the consequence, then, not of some overall purposeful effort to create a unity of design, but of necessity. It becomes apparent that the dry-stone walls change appreciably from one location to another, in the type of limestone used to build them, in how they are built, and in the patterns of field enclosures. An examination of local geological maps shows how frequently the rock types change from one place to the next. Obviously, the task of moving building

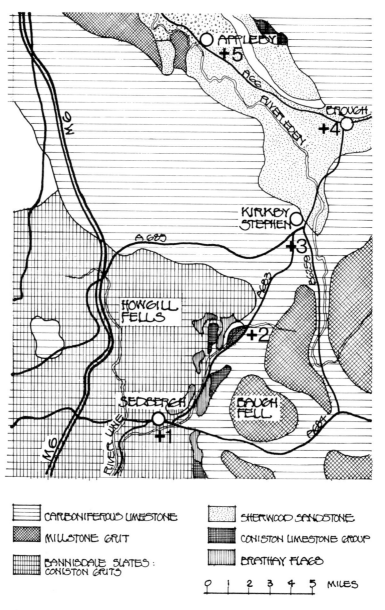

CARBONIFEROUS LIMESTONE

MILLSTONE GRIT

BANNISDALE SLATES : CONISTON GRITS

SHERWOOD SANDSTONE

CONISTON LIMESTONE GROUP

BRATHAY FLAGS

0 1 2 3 4 5 MILES

Geology revealed through culture. Four different types of surface limestone in a distance of a few kilometers creating four different types of wall construction and character. Shown here is a simplified surface geology map of a small section of the

Yorkshire dales, Sedbergh to Appleby (about 27 kilometers as the crow flies) with numbered locations where the photographs were taken. Geological information supplied by Dr. Barry Rickards, Department of Earth Sciences, Cambridge University.

Location 1.
This wall on the
south side of Sed-
bergh made of
Coniston Grits.
Boulders are ran-
dom, rounded, and
uneven, and the
wall shows the
classic dry-stone
walling technique
for this kind
of stone.

This modern dry-
stone walling being
built in the same
general location as
the wall shown
in location 1 has
a much neater
appearance.

Location 2. This wall, about 8 kilometers north of Sedbergh, is of Carboniferous limestone, which is quite different in character and color. The builders took their material from the immediate surroundings (note the surface boulders littering the foreground). Each place has a markedly different character because of the differences in surface rock. The walls on the far side of the valley are not wholly Carboniferous limestone. The rock beneath them is, in part, Coniston Grits, which is what the builders have used. The Dent fault, running along the line of the valley, brings Carboniferous limestone against Silurian.

Location 3.
Carboniferous
limestone wall.
This white, flat-
bedded, and un-
even rock is found
in the uplands
above the valley
floor south of
Kirby Stephen.

Location 4.
At Brough a few
kilometers north of
Kirby Stephen,
walling material
changes to pink
and brown Sher-
wood sandstone.
The flat-bedded
and squared mate-
rial makes for
more even-coursed
walling.

*Location 5.
In the valleys
where stone is well
below the surface,
hedges have re-
placed stone walls.
A markedly differ-
ent landscape is
thus created within
a very short
distance.*

stone manually from one place to another had to be reduced to a minimum. Thus the different types of stone required variations in construction.

In responding to local materials, therefore, building crafts developed distinctive regional forms. Once someone had devised a decorative feature, it would be copied by a neighbor. Farmers in southeast England vied with one another over the extent and handsomeness of the porches for corn barns, whereas in the Midlands, porches were almost disregarded.[8] Differences arising from fashion, materials, and craftsmanship are numerous and can be seen every-

where in the rural landscape. Carver comments that "the thatched roof has always posed the difficult problem of securing and sealing the thatch at the roof ridge. Elaborate solutions to this problem provided the opportunity for highly visible symbolic elements distinctive in each location."[9] In Dorset, for instance, variations of thatching are apparent from place to place; and roofs are often marked with a distinctive sculptural emblem in thatch on the end gable to identify the particular craftsman responsible. In other words, out of the basic requirement to provide shelter arose a symbolic cultural form.

Wherever one goes, one finds
similar forces shaping the landscape
of necessity. Portugal's southernmost
province, the Algarve, is a land overlain by Roman, Moorish, and Portuguese culture. To this day its
timeless architecture of rectangular
white buildings, ceramic tiled facades
and red tiled roofs; its squares,
churches, and fishing ports; its
people sitting in doorways, parading
in the streets, bargaining at the market, and unloading the day's catch in
the harbor join together to retain a
powerful sense of place.

Yet the power of the Algarve landscape itself pervades the towns, villages, and countryside. Its two
landscape regions are composed of
different types of rock. Sandstone
and limestone occur on the coast,
and granite is the material of the
mountains to the north. These give
rise to markedly different landscapes
and human responses to them. Red
sandstone formations, carved into extraordinary shapes by erosion, dominate the beaches and create complex
spatial environments. Inland, limestone gives rise to a man-made landscape of round topped cement walls,
stone filled and rendered, dividing
farms and private property. Limestone cobbles line the edges of the
roads. Limestone and marble pave
the urban squares and sidewalks.
Gracefully ornamented latticed
chimneys, originally made in ceramic
but today built of cement, adorn the

*Timeless expres-
sions of architec-
ture and place in
Portimeo, the Al-
garve, Portugal.*

The landscapes of necessity are overlain with Roman, Moorish, and Portuguese craft traditions. Using the materials available, regional expression varies sharply from one place to the next.

A doorway in the village of Alvor, the Algarve, Portugal. Ceramic tiled walls are typical of the region, and the doors and windows facing narrow streets insure continuous social dialogue and street activity.

roofs of houses and impart a distinctive skyline to towns and villages. An hour's drive north into the mountains reveals a quite different landscape. Now granite is the material for the rough stone walls and terraced farm buildings, some whitewashed some not, and for the pavers lining the edges of roads. An enormous quarry cut into the side of the hill scars the landscape but reveals much about the geology and underlying character of the region. Portugal's famous cork oak growing everywhere, its trunk and branches partially stripped of bark, tells one of the native vegetation and of local crafts on display at every roadside market.

Urban and Landscape Form and the Process of Time

The vernacular form of early building was a response to the practical problems of shelter and security. Decisions to build a city wall, to lay out a pattern of streets and squares, to locate settlement on the tops of hills or at the bend in a river, and how to use materials were functional and pragmatic, rather than the result of preconceived doctrine. Thus Cologne's thirteenth-century plan was symmetrical because there was no practical reason against it being so, not because symmetry was a virtue.[10] The tendency to romanticize the past and the remnants of other cultures embodied in winding cobbled streets,

piazzas, and ancient churches is typical of the average visitor who is able to wander through these places separated from the day-to-day struggle to survive.[11] As Carver has observed, for the medieval builders, building on hilltops was not a challenge to subdue nature, or to enjoy a view of the countryside, but to pose a challenge to their enemies and preserve precious valley land for farming.[12]

The picturesque qualities of form and texture of old towns, then, were largely the result of necessity—of having to deal with the realities of the time. In the absence of the technological means to flatten hillsides this inevitably led to what we see as a sensitive adaptation to site, and a forthright use of materials and building techniques. Consequently, the hilltowns and villages of the Tuscany countryside reflect the geological formations from which they were built, the materials for building streets, piazzas, and walls an outgrowth of the natural rock. It was the closest material to hand. We may also be fairly certain that if aluminum siding, concrete block, or corrugated iron had been available in the twelfth century it would have been used, the probability of which is demonstrated in the adaptations to problems of shelter in the shanty towns that surround most Third World cities. The availability of materials has always been a basic determinant of form, encouraging or preventing solutions for the built environment that, in the

Colle val d'Elsa, Italy. The symbiotic relationship of convenience and survival. The hilltowns of Tuscany enhance their sites where the landscape rather than built form dominates the scene.

absence of universally available materials, differ from place to place.

Most old towns, therefore, have a connection with the local environment, a sense of place, that appears remarkable to modern eyes. In the vernacular, architectural style is not an issue in the development of form, except in the detailed enrichment that comes from crafts traditions. This raises interesting issues about the questions of aesthetics and the preservation of old and beautiful places that we now regard as heritage landscapes, even though their functional basis for existence is long past. Seen from a distance, Italian hilltowns enhance their sites because landscape form rather than built form dominates the scene. From within, the view is of valleys and agricultural

land below—an intimate, and today, symbolic expression of the relationship between convenience and survival.

The transformation of the landscape itself over generations from one form to another is the consequence of cultural evolution that follows patterns similar to natural processes. Both are in constant transition, and what one sees today—a point in history—is not what one will see tomorrow. The underlying forces that have shaped the vernacular landscape teach us a great deal about the processes that made it the way it is, thus providing us with important lessons for design. At the same time it is apparent that the historical developments that have shaped the working landscapes that we admire for their sense of harmony and beauty today had, overall, as little to do with aesthetics as did the natural processes that made scenery.

A case in point is the Yorkshire dales, discussed earlier. The different patterns of walls forming field enclosures in the dales landscape are clearly apparent to the careful observer. First, the irregular pattern of crooked walls that may be seen near villages and hamlets creates a maze of small field enclosures. Second, a geometric, rectangular pattern of walls covers the valley bottoms and lower slopes of the hills and encloses larger fields up to twenty acres. Third, other, infrequent walls follow dead straight lines for miles up the higher ground and across the moors, parceling the land into long strips. These patterns represent three separate periods of walling that occurred in the sixteenth, the late eighteenth, and the nineteenth century, respectively. Each is "a monument to a social and political revolution of first magnitude as well as to the patience and skill of their builders."[13]

Sheep have for centuries been the mainstay of this region, and it is sheep that have turned the valleys and hillsides from the dense oak forest and scrubby woods of a thousand years ago to open meadow and scrubland. With the flourishing of the wool trade in the sixteenth century and the introduction of better breeds of sheep, much attention was given to improvements to pastures.

The enclosure of small fields that allowed for better fertilization and management of the land naturally followed. The irregular pattern of walled fields of usually a half-acre to one acre were created by independent farmers through consolidation and exchange of lands on the borders of old common fields and the open spaces around villages. The subdivision of common land in the eighteenth century demanded by the landed gentry was in response to new changes in agriculture. The result was land laid out in accordance with set standards, in straight walled and rectangular fields. These were expanded to include the larger upland landscape in the nineteenth century.[14] Thus, the essential character of this landscape is revealed through a combination of natural forces, generations of cultural history, and changing farming technology that occur both in the context of unorganized development and within a larger political framework. Changing times create changing landscapes.

The Imperatives of Climate

As different climatic regions are direct determinants of vegetation, animal communities, and land forms, so different climates also affect vernacular forms of building and landscape. In the absence of the technological means to ignore climate, built form has traditionally re-

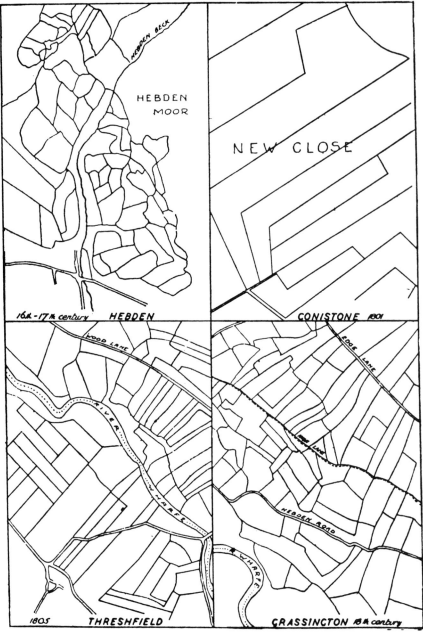

Field wall patterns in four adjacent townships in the north of England. These show the great variety in pattern and the steady development to straight-ruled walling under the Enclosure Commissioners, Coniston New Close walls and the Threshfield walls south of the river. All are to the same scale. (Arthur Raistrick)

sponded by developing direct and sometimes ingenious ways of moderating its effects. Possibly no other single factor has been as influential in creating a diversity of traditional forms and modes of living throughout the world. There is also little doubt that the most extreme climates have produced the greatest regional variations. The greater the constraints, the more powerful the incentive to find ways of providing human comfort within the means available. Conversely, the more benevolent the environment the greater the influence of culture on form.[15] In addition, although one normally thinks about the human response to climate in terms of the built environment, clearly the climatic determinants that have shaped the form of the nonbuilt rural landscape have been equally emphatic. The kinds of agriculture and patterns of the landscape that evolved from place to place up to the modern age of agribusiness were almost totally controlled by climate and soils. The long, hot, sunny summers, the slopes, soils, and microclimate of the wine producing areas of Europe have produced a characteristic and unmistakable terraced rural landscape. The carnation regions of the Alpes Maritimes in southern France have created another. The use of wind to draw water for irrigation has produced the characteristic windmill landscapes of Crete. Similarly, canal water levels were maintained at a

constant elevation by the famous windmills of the Dutch polders—examples of early landscapes of power.

At the coldest end of the scale, traditional Eskimo shelter was a perfect adaptation to nomadic life and to very low temperatures. The igloo that originated in and remained an invention of the Canadian Central Arctic was a house built of the only available material, snow. It resisted winds and minimized heat loss under the harshest conditions. It could be made in an hour or less and later abandoned when the tribe moved on. Building competitions that are held today in northern communities in Canada have shown that an expert builder can construct a snow house in twenty minutes and can then stand on it without it collapsing—a requirement of the competition.[16] In the western arctic the traditional building was the underground house, dug partially into the ground with the roof formed by timber or whalebone and covered by sod. Among the Plains Indians the "Tipi," or tent of buffalo hide stretched around a conical framework of poles, afforded excellent protection against the fierce winds and could be dismantled and transported as the tribe followed the migrations of the buffalo herd.

Building in the rigorous climate of the Swiss Alps involves a wider range of forms. Switzerland's geographic and ethnic variety over very small distances reinforces the strong perception of the country's regional di-

The imperatives of climate. Built form in extreme cold.

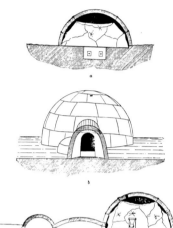

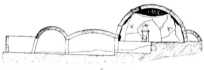

Snow houses of
Davis Strait.
(Boas, The Cen-
tral Eskimo)

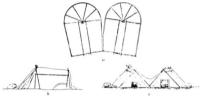

A diagram of a tupig *or tent from Inlet pond.*
(Boas, The Central Eskimo)

Double winter tent structure from
Cumberland Sound (Boas, The Central
Eskimo)

versity. Villages in the high mountain areas where cold winters prevail are closely packed, the houses drawn together for greater protection and warmth. Deep overhangs to cope with snow and rain, and storm windows to keep out the cold are typical features of Swiss Alpine architecture.

The evolution of the rural landscape in southern Ontario shows how dependent early settlements were on climate and how this dependence has changed over time. Rural roads were surveyed in the nineteenth century as part of the overall settlement grid planned by the British colonialists. When the road allowance was cleared in the early 1800s it resembled a tunnel through the forest with small clearings for farmsteads and fields on either side. Undisturbed snowfall onto the road made travel possible by one means: winter sleigh or sled. As settlement continued, however, the clearing of the forest resulted in windswept roads that made winter transportation by sleigh difficult. Subsequent roadside planting was intended primarily to catch the snow and allow passage of winter vehicles. Thus the road developed a new characteristic form that responded to the transportation modes of the time.[17] Today, with the change from sled to vehicular traffic, the function of trees to trap snow has become a liability

The steeply sloping roofs, stone walls, small windows, and narrow streets of Alpine village building respond to Switzerland's rigorous climate and its geographic and ethnic diversity over very small distances.

rather than an asset. This has resulted in the abandonment of planting, a preference for windswept roads, and the destruction of most of the old rural avenues.

At the other end of the scale, human response to heat has resulted in equally regional solutions to the problems of living in climatically difficult environments. The specially constructed wind scoops mounted on rooftops in the Lower Sind district of West Pakistan channeled the prevailing wind into every room and maintained more equitable temperatures than those outdoors. Houses in towns and villages were grouped close together to shade each other. The courtyard house and garden with its cooling plants and water have long

been associated with the hot, dry conditions of southern Spain.

The forms created from necessity in many Mediterranean towns today have created a new vernacular. Everywhere in the ancient town of Bodrum on Turkey's Aegean coast, for example, the flat rooftops of houses reflect a complex skyline of television aerials, modern solar panels for hot water, and awnings to protect rooftop living areas, or the streets below, from the sun. The new skyline of communications and electronic entertainment technologies expresses the social behavior, customs, and priorities that have always dictated the vernacular.

In agriculture, traditional methods of cultivating the North African des-

The imperatives of climate. Built form in warm climates.

The contemporary vernacular combines built form and style of living in warm climates. Solar heating panels, rooftop living terraces protected by awnings, and the ever-present electric power lines create a complex skyline.

Bodrum, Turkey. Time-honored traditions for shading streets from the sun in Mediterranean climates.

ert developed by the Nabatean tribesmen between the second century B.C. and the seventh century A.D. were highly sophisticated systems of agriculture based on the concept of maximizing the use of minimal rainfall (which averaged 50 to 100 millimeters per year). Known as run-off farming, it involved the construction of cisterns, dams, and barrages, and leveling and terracing of the land. The small farm units were irrigated from large water-catchment areas. These irrigation methods are being studied today in Jerusalem as a means of desert reclamation.[18] The moderate climate of the British Isles, with its gentle and persistent rainfall produces a characteristic lush green landscape that is the envy of all visitors. It also produces landscapes climatically adapted to pasture and sheep and cattle farming. The first book on farming published in England in 1523 explains that of all animals the sheep could be the most profitable. Sheep farming and the proceeds of wool largely account for the whole-sale rebuilding of rural England between 1526 and 1600. The famous wool towns of the Cotswolds are evidence of the prosperity of this period.[19]

The Landscape of Authority

There is, of course, always the temptation to think about all vernacular forms as the product of evolving practical decisions made incrementally over long periods of time and in the absence of overall planning or leadership. Other forces have, however, exerted themselves on the landscape at a scale made possible only by some central authority. It is important to understand the regional imperative in terms of the role of authority—the making of large scale, long-range decisions on the landscape that are imposed on the individual. J. B. Jackson has described this as the political landscape, one "which evolved partly out of experience, partly from design, to meet some of the needs of men and women in their political guise."[20] The walls, boundaries, highways, monuments, and public places he includes play a definitive role in both the rural and urban landscape and exist to insure order, security, and continuity. In short, they are the framework or organizing structure within which people live and work.

It can be argued that the absence of regional context is endemic to the landscape of authority. Only limitations of technology and transportation have forced the acknowledgment of local regional influences. But both the structured and informal landscape have always been present in history, and both continue to influence the evolution of the landscape today, the latter evolving within the organizing framework of the former.

The hydraulic civilizations that

historically have included China, India, and pre-Spanish Mexico were developed in arid regions that support land crops and in humid environments of aquatic cultivation. To make agriculture viable on a large scale required major organizational skills. The human investment in effort and money to develop flood-control systems, piping, ditching, land terracing, and so on, were beyond the capability of the individual peasant farmer. Water management had to be operated by a central authority with the power to establish new systems of division of labor and cooperation.[21] It created productive and distinctive landscapes that, while effecting major changes to the land, nonetheless established a regional identity by having to adapt its efforts to the natural forces of climate, physiography, hydrological cycles, human energy, and existing levels of technology.

The concept of the boundary, expressed in the rural landscape by hedges and walls dividing property and fields, had a similar political agenda of order and control. W. G. Hoskins has shown that the evolution of field enclosures continued at a varying pace in England in every century up to about 1730 and were the result of private agreements by land owners.[22] The feudal system with its built-in inefficiencies and poor soil conservation was incapable of supporting advances in agriculture. The emergence of mixed farm-

ing included improved cultivation and rotations, new crops and winter fodder, and improved strains of plants and animals. The new, rearranged landscape was also inspired by the need to reestablish timber resources and woodland for navy shipbuilding that had been depleted as a consequence of the Dutch wars and the 1666 fire of London.[23] It was, in effect, the transformation of an ancient open field system into the landscape that we see today. Although the imposition of this landscape on the rural poor resulted in enormous social cost, ecologically it was a process that saved much of the country from environmental ruin. The character of the transformed landscape underwent drastic change and was heartily disliked by many people at the time, but it reestablished a new and appropriate regional character that has become accepted over time as the essence of what a rural landscape should look like.

In countries where the cultural landscape has evolved over thousands of years, the landscape of authority has been a part of a long pattern of biological and social change involving landscape reconstruction. In North America, however, the imposition of planned order took place in a relatively short length of time, in an already established scientific age, and in a wilderness environment that was being exploited and destroyed for profit. Southern Ontario was laid out by survey long before any major occupation took place, unlike many other settlements in North America. The British imposed a rigid gridiron system on the majority of the lands, and by 1860 the entire southern part of Ontario had been surveyed and settled.[24]

Towns and cities mix the practical directness of vernacular building with the structure and overall imposition of form created by authority. Even in the most obviously vernacular towns that have grown "organically"—that is, without an apparent plan—there were, in most instances, organizing influences that dictated their overall shape. The control exerted by the political forces of authority over protective walls, streets, squares, and important buildings is an example. The influence of the grid, while adaptable to modification when site conditions require it, becomes a major organizing element where settlements are exported from their cultural places of origin to new non-native territory. The founding of settlements in the Americas and in the Far East was based on standards for layout that came from a system of surveying land of common origin.[25] Nonetheless, while Spanish and English layouts for new towns were generally similar, differences in culture and environment created very different places. In South America, streets oriented to avoid wind channeling, the arcading of building facades, the location and uses of piazzas, and the emphasis on churches and buildings

with a distinctive Spanish architecture contributed significantly to the making of very different urban environments.[26]

In early North America, the European tradition of dense, closely packed houses was rejected in favor of the "green country town," a concept that has continued to be an ideal of urban living today. The layout of Philadelphia, Mark Girouard notes, was modeled on the plans for the rebuilding of London after the 1666 fire (none of which, interestingly, were used) and the new city's squares were planned for recreation rather than markets and churches as was the custom in South American early towns.[27] So even though the gridiron plan and road system imposed by authority have traditionally ignored the land, their actual application to the land and differences in culture have often created a recognizable regional identity.

WHAT CAN usefully be learned from the vernacular forms of settlement that give a perspective on the essential ingredients of regional identity? The following seem to have some significance.

The vernacular forces that created functional towns and countrysides have close parallels to the processes of nature. They occur, for the most part, from an inherent drive to fill a niche, to seize an opportunity to flourish, or to enhance one's chances of survival or success. Consequently, vernacular landscapes, whether urban or rural, are the product of necessity and limitation. In other words, there are few options.

Regional identity is based on the limitations of technology, on limited options or choices to effect changes to the environment, or on one's ability to move freely from one place to another. As long as the need to work within the limits of environment and society exists, the sense of regional identity is maintained. Regional identity has to do with where one stays, where one's roots are, and consequently with where long-standing social traditions can develop. It is these traditions that provide the best opportunities for an investment in one's own place, in the community and the land.

The vernacular, both past and present, usually has worked within a variety of frameworks—those imposed by authority, by nature, or by both. While the framework imposed by authority usually has little regard for, or knowledge of, natural processes, the diverse human forces working within this framework enrich and diversify the physical and social environment. The biophysical patterns of the landscape, although altered by man, remain a dominant influence on the character of a region. The degree to which natural patterns are altered is a function of the availability and sophistication of the technology to hand. The sense of place is organic, changing with time. As new human needs and technologies exert themselves on the landscape, new landscapes emerge in response to the imperatives of changing conditions.

Both cultural and natural forces are responsible for creating a distinctive regional identity. Similar cultural forces produce similar forms in biophysically different landscapes. They also create regional differences where biophysical forces produce similar landscape forms.

Regional identity is perceived as an aesthetic primarily by visitors to places that have it. For the people who live there, it is their work and investment in the place that create it. For them the concepts of beauty and aesthetic have quite different meanings and in some cases such notions may be irrelevant. Aesthetic values for the most part have little to do with the creation of vernacular landscapes. Their perceived beauty and sense of fit is the consequence of the practical needs to solve problems of habitat and daily living. Beauty is the consequence of technological limitations which force adaptation to the land.

4 / Utopias

Ideals and Visions

Since the Garden of Eden, it has been an inherently human trait to search for an earthly paradise, for the kind of society that can reach an ideal state of stability, harmony, and happiness—in short, to seek utopia. Utopian writers sought different solutions to the social problems of their times, some in the pursuit of material property, some in austerity and simplicity, and others in technology and science. They were frequently associated with periods of social upheaval.[1] Lewis Mumford observes that a very large number of classical utopias were based on the concept of the self-sufficient city and its region and authoritarian discipline. Each was a closed society immune to human growth or change.[2] It will be useful to review briefly a few of them in order to understand how utopian thought has influenced perceptions and values, particularly as they affect the planning and design of our landscape today.

Plato's Utopia, described in his *Republic*, was an inquiry into what makes a just state. It was directly influenced by Sparta, a military democratic state that eventually defeated Athens in the Peloponnesian wars. It provided a classical precedent for ideas such as the control of property, money, education, and morals by the state, as well as sexual equality and the regulation of marriage and reproduction for the benefit of the state. For Plato, the ideal community was a city region, a city surrounded by enough land that could, based on a simple agricultural life, supply most of the inhabitants' food. Its citizens were divided into three classes—the common people, the soldiers, and the guardians—but the community was based on a common standard of living among the ruling and common people alike. Since wealth and poverty were seen to be harmful, neither existed in Plato's ideal city. Some two thousand years later, Thomas More's island *Utopia*, published in Latin in 1516, was also based on the city-region as the unit of political life with agriculture as its economic base. As in Plato's republic, all things were held in common, including property. Fifty-four cities and their surrounding land were based on the same plan: all streets were twenty feet broad; all private houses were exactly alike; and, people changed houses every ten years to avoid the problem of becoming too attached to property. Country farms, under the rule of a master and mistress, were worked by permanent rural people and by others sent from the city to work for two-year periods. Everyone worked six hours a day; family life was patriarchal, with married sons living in their father's house; clothing fashions were nonexistent. Bertrand Russell comments, "Life in More's Utopia, as in most others, would be intolerably dull. Diversity is essential to happiness and in Utopia there is hardly any."[3] Lack of diversity is, however, a defect of all planned social systems, actual as well as imaginary, although Mumford observes that the classical utopias all treated society as a whole. They were concerned with things that society should aim for in life.[4]

The Renaissance took these aims for granted and dealt with how man's scope for action might be broadened. Alberti's ideal city, for instance, was based on the view that the city should dominate the country—a shift of the balance of power.[5] Campanella's City of the Sun involved everyone in agriculture, with only the slow-witted working full time as laborers. And Roger Bacon, the forefather of the modern materialistic concept of progress, claimed in his treatise of 1627, *New Atlantis,* that science gives power over nature which leads to increased happiness.[6]

A fundamental change in the concept of utopias emerged during the eighteenth century. Rousseau viewed science as a force that brought moral ruin and society as merely a corruption of the original ideal state of nature. The rise of Romanticism was prompted by the growing sense of man's isolation from the natural world that resulted from industrialization.[7] The early industrial age and then the explosion of the nineteenth-century cities were seen as an era of brutality against the common people. The industrial utopias, unlike earlier ideal societies, were no longer concerned with values but with means and material well being.

In Europe and North America, people were being uprooted from their land and traditions for the factories of the cities. The appalling conditions resulting from virtual slave labor, slums, epidemics, and lack of sanitation stimulated the creation of a social reform movement of great purpose and moral force. It was from this movement that new ideals for public health, housing, and urban parks evolved. The two conditions—the changing countryside and the industrializing cities—created fundamental divisions between the two. The latter were seen as the epitome of evil and destroyed humanity, and many writers, Charles Dickens among them, deplored the corrupting social and physical conditions of the cities. The former became the basis for a series of utopias that looked to the land as a central condition of happiness and the regeneration of society, an ideal supported by reformers such as William Cobbett, Fergus O'Connor, and Robert Owen.[8]

Planning Utopias

The idea of regeneration through contact with the land and the need to find solutions for the worsening conditions of the cities were leading notions of planning utopias from the late-nineteenth century and up to the mid-twentieth century. Ebenezer

Howard, Frank Lloyd Wright, Le Corbusier, and others believed that what was required was a completely new and holistic approach to city building, one that would be based on social justice and equality. Interrelated revolutionary changes in urban design, politics, and economics had to take place if real solutions were ever to be found. Short-term solutions that tinkered with the system and its problems were, they felt, inadequate and ineffectual.

Ebenezer Howard's anti-urbanism was connected with the extremes of wealth and poverty he saw in the city. In his *Garden Cities of Tomorrow,* published in 1902, he turned to decentralization as a solution and concluded that large cities had no place in the society of the future.[9] Human society and the beauty of nature, he believed, were meant to be enjoyed together in communities of about thirty thousand based on small business and agriculture. His garden city, located in an agricultural belt of farms and woods, played an integral role in the economy of the city with two thousand farmers supplying the city with the bulk of its food, and the farmland itself preventing urban sprawl. Howard believed that the true remedy for capitalist oppression lay in democratic, cooperative organization. The garden city should build up the economic power of the workers themselves. The building of the first of Howard's two settlements,

Letchworth, was based, however, not on utopian hopes but on a solid business venture and good investment.[10] In the 1920s the garden-city movement lost its commitment to social change and became a city planning movement that, following the Second World War, led to the creation of the New Towns Act and the building of satellite cities around London.

Hygeia, the utopian plan for the healthy city proposed by the nineteenth-century visionary reformer Edwin Chadwick and pursued by the physician and sanitarian Benjamin Ward Richardson, was a visionary plan to enhance the physical well being of the urban population. It was created at a time when the physical and social conditions of the cities were so bad that disease and death were a part of everyday urban life, a fact of which Ebenezer Howard and others were acutely aware. Richardson saw a "community scrubbed clean as the kitchen of any Dutch housewife . . . a city of public baths, swimming pools, and gymnasia alongside its schools, libraries and lecture halls."[11] The population of Hygeia was limited to twenty-five persons per acre; every house had its own back garden and public buildings were surrounded by lawns and garden space. The paved streets were kept spotlessly clean and "debris was carried away beneath the surface."[12] With the passing of the Public Health Act in 1875 improvements in public health did occur in Britain. In cities like Manchester, the mortality rate, which had been 33 percent, dropped to 11.6 percent by 1940. In the United States a comparable decline was apparent, with less than ten deaths per thousand reported by 1950.[13]

Howard's view of decentralization as an answer to overturning capitalism and reversing industrial and economic trends became, with Frank Lloyd Wright's Broadacre City, a concept based on population dispersal as the prerequisite for individuality and democracy: "The dream of the free city to establish democracy on a firmer basis."[14] Wright hated the city and advocated a return to the land, where the citizenry could bail "out of the urban mantrap into the more natural life of the small town fruitfully expanded into the country."[15] In Broadacre City, decentralization reaches a point where the distinction between urban and rural form and style of living disappears and there is no recognizable center. Based entirely on the automobile and an extensive transportation network, it was a grid of homesteads, dispersed across the landscape in one-acre plots, and institutions, cultural centers, and factories.

Wright saw the personal vehicle as the means of providing a new mastery of time and space and on which a new kind of city could be built. Whereas the garden city was man on foot moving at three miles an hour and able to cross the city in fifteen minutes, Broadacre City was man in vehicle moving at sixty miles an hour and able to visit any of his thirty thousand neighbors in the same time—a radically different sense of space and location. But as Robert Fishman points out, there is little to suggest that Wright's view of a decentralized society would be a free or creative one. The northern United States in the first half of the nineteenth century was as rural, egalitarian, and decentralized as Wright's Broadacre City was intended to be, but it suffered from a deadening conformity of taste and opinion, precisely that tyranny of the majority which he deplored and attributed to the existing cities.[16]

The other great twentieth-century utopia, Le Corbusier's *La Ville Radieuse* (The Radiant City) was, like Broadacre City, an attempt to reconcile man with nature and transform the concept of the city. Unlike Wright, however, Le Corbusier's utopian notions of the totally planned society were to be found in the contemporary city itself. In 1920 he published *Vers une architecture* in which he outlined his views on the need for a modern architecture appropriate to the machine age. His famous pronouncement "A house is a machine to live in" was the expression of an architectural ideal, inspired by the functional design that produced the grain elevator, automobile, airplane, and steamship. Similarly, his grand view of the city was one that demanded the rebuilding of industrial society where man, nature, and the machine would be reconciled. The city of the future would be a radiant city of glass and steel skyscrapers set in great parklands. The plans express his concern for order and geometry and the symbol of planning over anarchic individualism: social order over discord with industry, housing and offices occupying separate sectors. The transportation system, designed for speed, became the central organizing principle that brought all the parts together. The central terminal and interchange were surrounded by skyscrapers, each building standing on its own in open space, thus freeing the ground for greenery.

The Radiant City symbolized the centralization of society and was dedicated to bureaucratic authority, to the ability of comprehensive planning to put the world in order. It was also a "Green City." Only 15 percent of the land was taken up by streets and buildings, the rest being devoted to pathways, trees, grass, playing fields, and the recreational needs of a full family life. Le Corbusier did not believe in piecemeal planning. He followed in the French urban tradition of Haussman which was based on the principle that only total demolition and transformation of the old urban texture could create a new urban order. He believed cities should be planned by experts who under-

stand the science of urbanism, and not left to the shifting views of the population.[17]

Scientific Utopias

In the 1960s a new cult of ideal cities surfaced, one that arose from the notion that developments in science and technology required a new vision for an urban society. Among these, solutions for the environmental and social problems of humanity were put forward, notably in Soleri's principles of archology (ecological architecture). Archology replaced urban sprawl and suburbia with huge concentrated megastructure cities housing hundreds of thousands of people in a self-supporting environment that married biology and high technology. Its guiding principle was that by concentrating people in one place the rest of the natural landscape could remain undisturbed.

Buckminster Fuller's concept of spaceship earth was a vision of the world as a limited entity which espoused high technology, design, and the conservation of natural resources. Fuller stressed that resources should be used with greater efficiency, and with his invention of the geodesic dome he foresaw vast air-conditioned spaces under which whole cities could be housed, creating ideal urban environments for the future. New York and Toronto were among the major cities for which he made such proposals, and the theme has been suggested for many a northern settlement as a solution to problems of living in hostile climates.

Eutopia versus Utopia: Issues of the Regional Imperative

When Patrick Geddes coined the word *Eutopia,* meaning "good place," in his address to the Sociological Society in July 1904, he proved too much for some of the intellectuals of London.[18] In comparing it with the commonly understood *Utopia* coined by Thomas More, a word derived from the Greek "no place," he summed up a fundamental tenet of the regional imperative: that it makes sense to design with the forms and cultural and ecological processes already present in a location rather than to force an idealized, preconceived plan upon a site. *Eutopia* is assured when culture and ecology become part of design thinking. *Utopia* is the consequence of ignoring them. Indeed, the tendency for conscious design to perpetuate the conditions of *no place* is at least as strong as the present day forces that are shaping cities and the regional landscape.

From an environmental perspective, there are common threads that are woven through most utopian thought. By their very nature utopias are the product of individual intellectual ideas, and invoke an authoritarian, static concept of life and nature that is immune to change. Mumford comments that each utopia was a closed society for the prevention of human growth; and the more the institutions of utopia succeeded in stamping the minds of its members the less possibility existed for furthering creative and purposeful change.[19] A great many have been based on connections with the land, the rural life and agricultural production, with the notion that man and nature should be reunited.

But in these nature is seen as a humanized benign backdrop to the search for social ideals of justice and the good life; it is a nature that is tractable to human needs and desires. The notions of order, equality, and happiness emanating from idealistic visions of total change have little to do with the nature of the processes that govern life. Human ideologies and the laws of nature are inherently incompatible. They are, in every sense of the word, No Place, the antithesis of regional diversity and identity, conceptually the imposition of a static authoritarian view on biophysical and human processes. As Milan Simecka observes, "All the indications are that Utopias are nothing but the instrument of a historic deception . . . the question arises whether Utopias actually contain the seeds of their eventual downfall, whether any and every attempt to put them into practice will eventually produce a caricature of their original intentions."[20] Divorced from its environmental implications, a concept such as Hygeia and the healthy city can be seen to have brought about the deterioration rather than the enhancement of health in its larger sense. In the nineteenth century, for instance, advances in sanitation to combat disease as the cities grew larger and more crowded were based largely on the development of piped water supply, sewers, and sewage treatment technologies. But with these technological advances, clearly necessary to human health in the cities, came the deterioration of the larger environment as the wastes from the city polluted rivers, land, and air. Also inherent were the seeds of environmental attitudes that continue to plague us today: waste and consumerism; misuse of resources; the throw-away society; and an unwillingness to make the connection between social or economic benefit and environmental cost. The benefits of sanitary health, of well-drained and clean streets are paid for in environmental costs in the larger landscape.

It can be argued, therefore, that nothing could be more fatal to society than to achieve its utopian ideals. Yet some proposals for planned cities, particularly those of the first half of the twentieth century, have had enormous influence on the contemporary landscape and in the shaping of the underlying percep-

tions and assumptions of modern design. These proposals were the consequence of a deep-seated belief that rebuilding the city is the first and most important step in transforming society.[21] Nothing less than total change based on doctrines of social betterment and aesthetic form could solve the perceived problems of the city. The utopian cities proposed by Wright, Le Corbusier, and others were plainly unworkable. Yet their ideas survive in the built landscape of today, as well as in the tradition of intellectual dogma passed on by generations of planners, designers, and decision makers who see their role as determining the future of society. Jane Jacobs comments that "[bankers] have gotten their theories [about cities] from the same intellectual sources as the planners. Bankers and government administrative officials who guarantee mortgages do not invent planning theories nor, surprisingly, even economic doctrine about cities. They are enlightened nowadays, as they pick up their ideas from idealists, a generation late."[22]

To anyone traveling through Europe and North America the results of what society does with utopian dreams are plain. One finds Corbusian landscapes in the oppressive high-rise residential developments typical of centralized regimes where the state provides for its citizens (suburban Paris or Warsaw) as well as in Western cities, where the spontaneous social interactions that come from street-level living are no longer possible. Wrightian landscapes are found in the free-enterprise suburbs of the United States, Canada, and Western Europe. The landscapes of Howard exist in the new towns of Britain, which were originally intended to stop London's growth by siphoning off excess population into satellite communities. In 1948 it was assumed that the introduction of planning would solve the two great urban problems: decay in the city core and sprawl at the edges.[23] This latter assumption has proven to be incorrect as London has continued to grow.

Such idealism can also be found in the satellite towns around Stockholm. The last word in planning theory in vogue in the 1950s and 1960s, they included the separation of people from traffic, the continuous greenway system, the shift of social activity away from the street to the linked landscaped space at the back. Like my contemporaries, I was excited and influenced by these ideas and anxious to study them in place. But the reality seems to inspire the tedium, lack of spontaneity, and a sense of sameness that such controlled design experiments so often carry with them. Yet as Christian Norberg-Schulz has observed, these theories have resulted from a wish to making man's environment better.[24]

The Persistence of Doctrine

The issue of utopia, however, still persists in the 1980s. Through exhaustive and scientific study of high-rise housing in Britain, Alice Coleman has shown how the postwar visions of ideal housing environments in fact have resulted in squalor, vandalism, and social breakdown—the very opposite of what they were intended to be. On trial is the utopian view, through the medium of design control, about how people should live.[25] As Coleman observes, such deeply embedded ideals have had a lasting impact on the social and physical environment. With the perspective that we enjoy today and the hindsight it affords, we can understand why Daniel Burnham's famous exhortation to "Make no little plans for they have no power to stir the minds of men" must be one of the most ecologically innocent statements made in the twentieth century.

In 1961 Jane Jacobs wrote, "Cities are an immense laboratory for trial and error, failure and success in city building and city design. This is the laboratory in which city planning should have been learning and forming and testing its theories. Instead the practitioners and teachers of this discipline have ignored the study of success and failure in real life, have been incurious about the reasons for unexpected success, and are guided instead by principles derived from the behavior and appearance of towns, suburbs, tuberculosis sanatoria, fairs and imaginary dream cities—from anything but cities themselves."[26] Jane Jacobs talks about cities as they work in real life rather than about theories and predetermined ideas about how they should. In this she strikes at the heart of a problem that persists in contemporary urban design: that successful, enriching, and diverse life is a product not of predetermined doctrine, but of vernacular forces that work because they have come about through trial and error in response to the constraints of environment and a thousand small decisions made in the interest of practical necessity. Inherent in her view is the fact that the interrelationships and interactions of urban life are far too complex to be created in some magnificent, comprehensive plan emanating from an individual mind.

This view of urban social forces evolving over time has essential and persuasive parallels with ecological processes that shape regional identity, and it highlights the attitudes of environmental planners and designers toward managing nature. Design carries with it a burden of values and inner vision about what is good for the environment and for people. Our urban design tradition has been dominated by a view of nature as humanized landscape, a view aided and abetted by horticultural science and the aesthetic that has evolved from it.

Landscape in cities, and indeed wherever design is imposed on the land, represents nature under control. The extraordinary natural diversity that is expressed in most cities in their varied climates and geological formations, their hills and river valleys, their native and emerging plant and animal communities—those elements that tie a place to its region, that give it identity—is subjugated to the utopian vision of the perfect, unchanging cultivated landscape of turf, ornamental trees, and herbaceous borders. In effect, design ignores regional identity.

The emphasis of abstract design on paper reinforces the dichotomy between the idealized plan (what the designer has in mind) and the need to observe the nature of the place (what the landscape is and how it works). Evidence of this is everywhere. The creation of benign outdoor microclimates is ignored in the design of urban spaces, parks, and plazas in favor of the air-conditioned interior. The geodesic dome has persisted since the 1960s as the utopian "high-tech" solution to the planning of settlements in cold climates. Adult-conceived playground design continues to be imposed on the inherent creativity of children, when all direct observation tells us that children do not play how and where they are supposed to. There is the myth that the more recreational open space one has in planned development the better and the happier people will be.

The wholesale imposition of "creative" play equipment on children in Iqaluit, Baffin Island. The preconceived adult vision of how children should play is one of the causes of no-place. This scene shows what the kids think of prefabricated play structures (in the background). With the eminent sense and inherent creativity of the young, they ignore them for the more useful material that nature provides.

Ignored is the readily observable fact that quantity has little to do with the social health and identity of neighborhoods. Deeply ingrained in our consciousness and image of urban planning is the notion that the exclusive function of open space is to provide recreation and visual amenity. Urban planning is based on the assumption of a continuing evolution toward a leisure future for humanity, where work, as it applies to land and investment in nature, has less and less relevance or purpose to the city. This view has its foundation in the nineteenth-century urban park and in subsequent grand plans for the reform of cities and changing ethics of work and play. Today's park systems are often unrelated to the productivity of the land as a place of work, with its ecological role in the maintenance of health, or with its function in the creation of self-sustaining urban places.

Image versus Reality

Like aesthetic perception, which is the individual response to what is beautiful and what is not, utopias are the product of individual minds seeking idealized solutions to the perceived problems of society. It is part of human nature. Yet even though Hygeia did lead to more healthy cities and though the green park did provide much needed contrast to the city's congestion, every person has different ideals and different views on how these goals might be achieved. Therefore, one person's vision for society imposed on the collective community is not only wrong but also unworkable for the same reason that makes it impossible to control the forces of nature for some predetermined design purpose or form.

Idealism and predetermined images are very much in evidence in our perceptions of the landscape, and they inhibit a clear observation and understanding of the world around us. This attitude was amply demonstrated in observations made over two decades of field courses on stream hydrology at the University of Manitoba by the distinguished Canadian hydrologist and teacher, R. W. Newbury.[27] Over this period Newbury took four groups of undergraduate students with different academic backgrounds to survey the stream dynamics and natural processes of a small watercourse in Riding Mountain National Park. As part of the exercise he asked individuals in each group to map their observations of the stream and its environs in a way that would be understandable to the others in the group. The accompanying sketches show the results of this effort.

The biologists saw the stream channel primarily in terms of aquatic habitats and its "organic form." They recognized the curvilinear form of the stream course, the cutting of banks as it changed directions, the deposition of eroded materials, exposed shale, wooded slopes, and flood plains. The behavior of the stream itself, however, was generally ignored. The engineers saw the stream as an almost straight manmade channel, flowing smoothly down its course and uninterrupted by obstructions. Its surrounding environs were areas of erosion and slump, and all were drawn with an engineering certainty and precision that defined problem areas that had to be corrected. The landscape students, with no training in engineering or biology, created fine artistic renderings (often the only sketches done in color) that showed stream deposits, flood plains, and a thin, wavy stream line that revealed nothing of the watercourse's behavior or its influence on its surroundings. The technicians, with little artistic skill or environmental training drew, however, a functional diagram of the stream as it actually flowed. They showed how it reacts to obstructions, flowing around large and small boulders, channeling through small openings, and how it is forced into alternate channels in times of spate. They illustrated crudely but simply

The biologists saw the stream as a series of habitats.

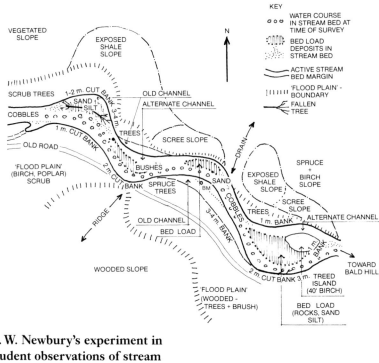

R. W. Newbury's experiment in student observations of stream dynamics. These examples, redrawn from information supplied by Newbury, show how the inner visions of differently trained people could direct observation and understanding of the place.

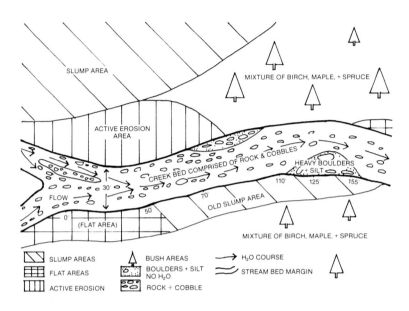

SLUMP AREA

MIXTURE OF BIRCH, MAPLE, + SPRUCE

ACTIVE EROSION
AREA

CREEK BED COMPRISED OF ROCK & COBBLES

HEAVY BOULDERS
+ SILT

30'

70'

110'   125'   155'

FLOW

50'

OLD SLUMP AREA

0'   (FLAT AREA)

MIXTURE OF BIRCH, MAPLE, + SPRUCE

SLUMP AREAS   BUSH AREAS   H₂O COURSE

FLAT AREAS   BOULDERS + SILT
                     NO H₂O   STREAM BED MARGIN

ACTIVE EROSION   ROCK + COBBLE

The engineers saw the stream as a series of problems to be put right.

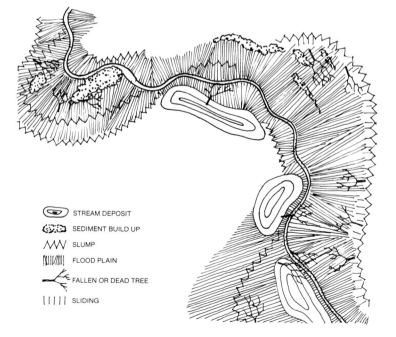

STREAM DEPOSIT

SEDIMENT BUILD UP

SLUMP

FLOOD PLAIN

FALLEN OR DEAD TREE

SLIDING

The landscape architects saw very little but did a nice drawing— in color.

The technicians provided a rough but accurate description of how the stream behaved.

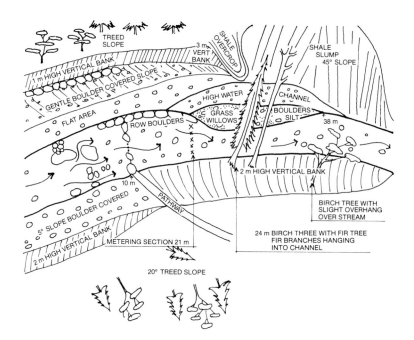

the various conditions surrounding the stream in a way that would explain to others what was there.

The drawings illustrate how observations of a real place can vary to a marked degree with whoever is doing the observing and how strongly influenced the observers are by their individual backgrounds. It is curious, too, that the environmental designers' sketches were both the most artistically competent and the least environmentally revealing of the place, while the technicians made the most accurate and the least biased observations of what lay before them. This example of how environments are perceived in the modern world of

communications is an illustration of the difficulty designers, and indeed people in general, have in observing the real world around them and in developing conceptual ideas that are based on what is there rather than what is imagined, or wished for.

It also shows, in miniature, the larger problems that affect the regional environment as a whole. Decisions to divert whole river systems, for instance, are made on the basis of images of the landscape derived from maps rather than direct observation, images that are far removed from the reality of how river systems actually behave. The impoundment and diversion of the Churchill River in

The stream as it actually is.

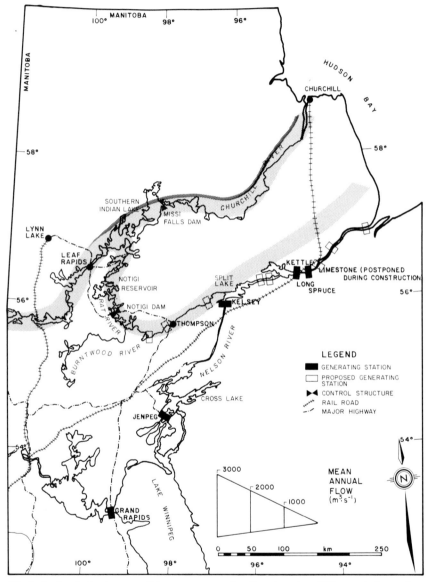

The hydroelectric development of the
Churchill and Nelson rivers, indicating the
altered flow regime of the rivers. The dark
tone indicates the relative magnitude of the
lower Churchill River discharge remaining
after diversion; the middle tone indicates the
portion of Churchill River discharge diverted
at Southern Indian Lake; the light tone
indicates the Nelson River discharge. (From
R. W. Newbury et al.)

northern Manitoba is one example. In 1976 the longest and one of the largest river diversions in northern Canada was commissioned to create hydroelectric energy for southern markets in Canada and the United States. Over 80 percent of the Churchill River (958 cubic meters per second) was diverted southward through 300 kilometers of tributary channels to the Nelson River in northern Manitoba. The environmental, economic, and social effects of this were immediate and disastrous as permafrost lands disintegrated under floodwaters, fisheries collapsed, and high mercury levels were discovered in the fish and native people of the river valleys. Ten years later, negotiations for compensation, land replacement, and native rights were still in progress.

The background that led to the decision to proceed with the project is instructive. Detailed hydrometric and design studies were undertaken between 1946 and 1968 and were followed by environmental impact studies between 1966 and 1975. As the areas affected by the diversion were far more extensive than the sites of the hydraulic works, it was apparent that only two options were available. The studies had to be either abandoned as impossible to accomplish in the limited time available, or they had to be carried out superficially in an "idealized" landscape. The latter course was followed and,

predictably, the idealization concealed impacts of such magnitude that the environmental instabilities created along more than 10,000 kilometers of lake shores and river banks were essentially beyond control.[28] Yet on a map, the diversion between the basins of the Churchill and Nelson rivers is only 30 kilometers long.

There is no question that twentieth-century communications have radically altered our ideas about how we relate to the world around us. The depersonalization of direct experience by the media and information systems places sensory barriers between the designer and the environment that inhibit the visualization and understanding of landscapes as real places. The increasing tendency to become disconnected from direct interaction with the environment through personal experience often has to do with the sheer scale and complexity of the problems that face the regional designer.

The computer, for example, has provided a tool for processing large amounts of information in the analysis of regional and local landscapes. Yet it is tempting to think about information systems as ends in themselves, rather than as bases for environmental insights and ideas. Theodore Roszak comments, "Very nearly the whole of modern science has been generated out of a small collection of metaphysical, even aesthetic ideas, such as [Newton's] 'Nature is governed by universal laws,' or [Descartes's] 'The universe consists of matter in motion.' None of these ideas is a conclusion arrived at by scientific research; none of them is the result of processing information. Rather, they are premises that make scientific research possible and lead to the discovery of confirming data."[29] Similarly, the direct experience of landscape that uses all the human senses in understanding its processes and in marveling at its complexity and beauty has little to do with information technologies.

Few, however, would argue against the absolute need for regional landscape planning. Information technologies are essential aids to the evaluation of complex data and land-use options that would otherwise be enormously time consuming. The danger lies in their potential to diminish our sense of understanding of the place, of environmental awareness and values.

The Regional Garden: Oasis as No Place

The age-old concept of the oasis has evolved from the notion of the ultimate utopia—the garden. At a small scale the garden symbolizes the cultural idea of a fertile and beautiful place in an arid or hostile landscape. As such it has added to the environmental and cultural diversity of many places. The urge to expand this ideal

to include entire regions, in arid landscapes like Southern California or Arizona, has been achieved at enormous environmental costs in the withdrawal of groundwater, the transfer of water supplies from distant river systems, and the consequent loss of the essential identity of the natural region.

The phenomenal growth of the wealthy sunbelt regions has been driven by the extraordinary richness and diversity of their landscape, the irresistible draw of a warm climate, and the mystique of constant sunshine, an idyllic environment and opportunity for "the good life." By the early 1900s Southern California had acquired a reputation as a Shangri-la and paradise on earth for the infirm, the afflicted, and the elderly.[30] San Francisco and the Bay Area, and Los Angeles and San Diego are situated in a dry Mediterranean climate and biotic region with average annual rainfalls of twenty-one inches and twelve inches respectively.[31] In fact, according to the 1980 census, California's urban areas accounted for more than 90 percent of the total population of the state, most of its 158,693 square miles being virtually empty. The greater proportion of the urban population is clustered in areas of dry grassland, chaparral, and coast scrublands.[32] The lure of an idyllic climate and beautiful scenery has spawned the growth of low-density housing, industrial developments, tourist attractions, and a vast agricul-

Transforming the desert into the Garden of Eden. The utopian dream realized at the expense of nature. The contrast between the lush plants imported from high rainfall regions elsewhere and the native vegetation of the hills beyond is a powerful expression of unsustainable development and lack of connection with the place: Florida in California.

tural industry, all of which have created an insatiable demand for water. The statewide system of aqueducts that carry water from where it is plentiful to where it is scarce has, at the same time, had irreversible impact on the environments from which it is drawn.

These sunbelt regions represent the utopian dream realized at the expense of nature. Nearly 75 percent of natural stream flow occurs in the northern third of the state, whereas 80 percent of annual water consumption takes place in the southern, water-poor two-thirds.[33] It has been calculated that at the current rate of demand, the development potential of surface water can leave little hedge against other prolonged droughts such as occurred in the mid 1970s. The increasing use of irrigation for crops from rivers and groundwater diversions accounts for some 85 percent of all water consumption with the remainder going to urban and commercial uses.[34] Almost half of domestic and commercial water use in the urban areas is for landscape irrigation, which rises to 75 percent in hotter, lower density communities.[35] The placelessness of this contrived human environment symbolizes the imbalance between the vision of a utopian Garden of Eden and the ability of the environment to sustain the ideal. Looking over the valleys to the surrounding hills in the Los Angeles urban region, one is struck by the stark contrast between the two and the clear indication of urban plenty in a resource-poor environment. The hills, browned by summer heat and drought but turning green in spring, represent the sustainable native vegetation that covers this semi-arid region. The valleys, perpetually green and lush with lawns, shrubs, and trees, represent the artificial, urbanized landscape sustained by massive irrigation. Most of the vegetation used in urban areas in Southern California originates in plant communities in other parts of the world where rainfall is higher. Dependent on artificial conditions to sustain them, they create an opulent landscape, "an unnatural quality that pervades the environment, and probably the psyche of Southern California."[36]

Solutions to the constant conflict between where urban growth wants to occur and the sustainability of the regional resources that make this possible are to be found in values that focus on a harmonious relationship between urban communities and the climatic and biophysical settings they depend on. The implication for

urban design of this view involves an understanding of the natural and human development processes shaping the local region. A study on the conservation of water in a semi-arid landscape in Southern California showed that outdoor water use in the City of Industry (part of the Los Angeles urban region) amounts to about 40 percent of the total water supply being used in the city. Of this, 60 percent is used to irrigate turf, 30 percent to irrigate trees and shrubs, and the remaining 10 percent is used to wash streets and cars and fill pools.[37] The study outlines strategies for conserving water including a range of urban plants adapted to low water consumption, reusing wastewater, low-volume irrigation, and capturing more urban and natural run-off. A combination of conservation measures, it is calculated, would reduce the use of water by 50 percent.[38] The implications for reconnecting urban areas to their local environments flow from such strategies. In the face of consumer values, however, good intentions, rational common sense, or environmental ethics have little impact on change unless the perception of absolute necessity dictates a departure from existing development patterns. Necessity drives the regional imperative on which design can capitalize. This is well demonstrated in the urban vernacular landscape of Tucson, Arizona.

Tucson and the Imperative of Water

Tucson's desert climate with its reliable sunshine, blue sky and dry air has, like the Southern Californian cities, attracted people seeking sunshine and health. Set in one of the most spectacular desert landscapes in North America, its intense and strongly identifiable character lies in its diverse desert-adapted plant communities that include the stately saguaro cactus, unique to the American continent. Like California, the postwar migrations of newcomers to the Far West brought hordes of people to the desert. They were followed by industry and agricultural development that could only be sustained by tapping groundwater from natural reservoirs beneath the dry land surface. In Arizona the same perceptions about the environment have prevailed as in the Californian regions: the expectation that a dry, hot desert environment can be transformed into a lush, high rainfall one, while still enjoying the advantages of the dry air, constant sunshine, warmth outdoors and air conditioning indoors—a prescription for environmental catastrophe.

In 1959 I visited Tucson and found such a landscape, its streets, boulevards, and public parks irrigated to support emerald green lawns and lush trees and shrubs totally alien to the plant communities of the surrounding desert. The city might have been dropped by helicopter from some exotic far away place: Florida in Arizona. I visited Tucson again in 1987 and found an urban landscape transformed. Its once green public spaces and private gardens had been replaced with a landscape of cactus, giant saguaro, prickly pear, pala verdi trees, jumping cholla, mesquite, and many other desert species set in the gravel and rocky soils typical of the native desert environment. The city, seen against the bare mountains that enclose the valley, suddenly had an intense sense of belonging to the landscape that surrounds it. The only major difference between poorer and more wealthy neighborhoods lay in the relative size, density, and splendor of the desert landscape designs. Mowers were nowhere to be seen, an extrarodinary and piognant reminder that this was indeed a different place.

The change from a landscape alien to its region to one that reflects it came about through the urgency to conserve Tucson's only source of water—underground aquifers. This necessity brought about the introduction of legislation that requires groundwater withdrawal to be balanced by natural replenishment. By contrast, Phoenix's water is derived from three large lakes that collect melted snow from the White Mountains via the Salt River. A plentiful water supply derived from water reclamation and retention from far away has perpetuated its artificial high-

Tucson today. The earlier, high-rainfall landscape has been replaced by one that utilizes native desert vegetation. The city, consequently, has a powerful sense of belonging to its surrounding region. Here, a median on a suburban main street.

A typical suburban property, its front yard landscaped with Saguaro cactus.

*Landscape planting
at the University
of Arizona.*

The Tucson fringe. Houses are set in the matrix of the native landscape, insuring continuity and creating a definite sense of place.

rainfall urban landscape, one without context in its region. A comparison between the average daily water consumption per person in Phoenix (some 260 gallons [984 liters]) and Tucson (some 155 gallons [587 liters]) clearly expresses the different priorities of the two cities in maintaining a level of sustainability with the natural environment. Tucson schoolchildren study a comprehensive program on water conservation called "A Sense of Water," created by the Southern Arizona Water Resources Association and one of the Tuscon school districts. Tucson residents are bombarded with television commercials and newspaper advertisements to reduce water use in summer. To encourage economy,

water rates increase as water use by residents increases.[39] Between 1976 and 1986 per capita use dropped from 205 gallons (776 liters) per person per day to 155 gallons (587 liters).

Unlike Phoenix, Tucson is one of the largest metropolitan regions in the United States totally dependent on groundwater for municipal, industrial, and agricultural supplies. In 1980 irrigation-based agriculture consumed 54 percent of the total amount used, with municipal and industrial consumers each using 24 percent. Five times more groundwater was depleted than replenished.[40] The water table was found to be declining throughout the region at rates varying between a third of a meter and two

meters a year. Projections showed
that Tucson's population would triple
over the next fifty years from
500,000 people in 1987 to 1.6 million in 2037.[41] It was against this
background that in 1980 the Arizona
legislature established the Tucson
Active Management Area (covering
about 4,500 square miles [11,520
square kilometers]) with powers to
institute and enforce groundwater
management over the entire hydro-
logic basin. Its mandate was to re-
verse the trend of water depletion
that had existed for a long time. It
was recognized that water conservation was central to the city's continued survival and that the rate of
withdrawal had, in a variety of ways,
to be matched by the rate of replenishment. There were a number of
management tactics used.

The Central Arizona Project. This
strategy involved the diversion of the
Colorado River to the west to bring
water to communities in central Arizona by 1991 and to provide Tucson
with two-thirds of its total supply
of water. Its purpose is to reduce
groundwater pumping and achieve a
long-term balance between with-
drawals and replenishment.[42]

Use and Control of Floodwaters.
Streams and rivers are dry and are
in spate only when heavy rains occur.
The most significant, positive contribution of floodwaters is as the sole
natural source of recharging ground-

water for the management basin. The
district concentrated on a program to
resolve the most crucial problems of
flooding, including integrating the
transportation and flood-control systems, removing existing residential
development from the floodways of
major rivers, channeling streams, and
constructing a series of detention
and retention basins that recharge
water into the aquifer.

Recycled Effluent Water. The effluent
from wastewater treatment plants has
been used in Tucson as irrigation water for golf courses. A metropolitan
effluent delivery system, when completed, would deliver 30 million gallons (113.5 million liters) effluent
each day for irrigation and groundwater recharge.[43]

Water Quality. For more than a hundred years wastes from septic systems and industry have been dumped
on the surface of the land, some of
which have contaminated the groundwater on which people depend. To
counteract this problem, the Pima
County Department of Wastewater
Management initiated a program to
control industrial waste discharges to
county sewers. This substantially reduced industrial wastes arriving at
their treatment plants.[44]

Agriculture. The single largest water
user in the Tucson Management
Area, agriculture is being phased out
in the Avra Valley. Acreage under cul-

tivation in 1952 was 60,000 acres (24,300 hectares).[45] By 1985, 20,000 acres (8,100 hectares) had been acquired with an additional 9,000 acres being planned to protect some 4.5 million acre-feet of groundwater for future use.[46]

THE DIFFERENCES between the two Arizona cities are symbolized by the stark contrasts between a sense of urgent conservation measures in one place and profligate use of water in the other. In Phoenix lakeside lots are considered a necessary part of life. In Tucson only one lake development was attempted and it went bankrupt.[47] In a land where water is a sign of wealth, the element of choice dictates issues of conservation. In Tucson persuading the public about the evils of an "oasis mentality" was not difficult since there has never been an abundance of flowing surface water. Unlike Phoenix where surface water abounds, people accepted the desert because there was no real choice. "Given that choice [they] did what anyone on earth would do living in the desert—find or create a large pool of water and put palm trees around it. The first large civilizations on this planet evolved in deserts and were based on tapping rivers and creating irrigation canal systems to transform the desert into Eden."[48]

Basic changes in attitude away from notions of wealth and abundance and toward conservation are inspired less by environmental ethics than by necessity. The history of settlement and sophisticated technology for ignoring nature in a naturally bounteous land militate against the conserver ethic, unless it is backed by the imperative of need.

5 / The Urban Region and the Loss of Identity

In March 1987 an article in the Toronto *Globe and Mail* reported that French gourmets were up in arms over the degradation of their favorite form of culture. "French cooking is in danger of going to hell in a handcart as fast as you can say 'Donnez moi un Big Mac avec frites et un milkshake au chocolat to go.'"[1] There are few more powerful indicators of what is happening to contemporary society than food. From the utopian heights of French cuisine or "the relatively humble (although delicious) heights of apple pie, maple syrup and cheddar cheese, we [in North America] have fallen to the sub-basement, gastronomically speaking."[2] Most people might agree that the products of mass culture are not only homogenized milk or pasteurized industrial cheeses, which the French decry, but every facet of contemporary life and experience.

As I suggested in chapter 3, the visual character of preindustrial landscapes was shaped by necessity. There was no alternative but to accept the limitations imposed by nature, culture, and technology. The differences between one place and another, the sense of belonging, of being rooted to a particular location, have traditionally been achieved because there were few alternative options available. The overall form of vernacular settlements was determined by the constraints of the land and climate, by building materials, and by the social and historic forces that were unique to each place and time. The lack of choices forced a recognition of regional imperatives. The apparent shift away from what is distinctive to what is similar in the contemporary world is the consequence of the complex social, economic, and technological changes that have occurred with increasing rapidity since the industrial revolution. In this chapter I examine some of the changes affecting the landscape and man's relationship to nature that are the consequence of the shift from rural to urban societies in the industrialized countries.

Urbanism without End

In chapter 1 I examined the images of identity in the central areas of the city, the established foci where people tend to gravitate and where the identity of the place is clear. In the expanding suburbs, however, other forces take over. Look beyond the center of Istanbul, London, Paris, or indeed almost any thriving metropolis and another quite different environment is revealed—one that is not featured in the guidebooks. If it were possible to transport a visitor on a magic carpet around the world and set him down in the suburbs of Toronto, Bournemouth, or Chicago, it is quite likely that he would have difficulty knowing where he was. Ask him to identify whether he is looking at the skyline of Calgary, Houston,

The expanding city. Are these the suburbs of Toronto? Chicago? Milan?

They are none of these. The place is Istanbul.

This is downtown Hong Kong, although few might recognize it. Without its mountain setting it becomes any North American city—universal form created from economic forces. Compare this image with the view of Hong Kong on page 11.

Vancouver, or Los Angeles, and he would probably be equally baffled.

It is a truism to say that urban places all over the world seem to be suffering the same kind of homogenizing fate as we find in other facets of contemporary life: the fried chicken franchise (universal food), musak and AM radio (homogenized sound), theme parks (consumer entertainment), language and Newspeak (a nanny is a staff person who interfaces on a habitual basis with the children). Urban growth over the past two hundred years, but especially since the Second World War, has been remarkable for its scale and speed, and since the nineteenth century it has been the dominant force affecting people and the environ-

ment. The new age has created a landscape of transportation systems, tower blocks, freeways, vacant lands, and suburbs that are continually expanding and swallowing up the rural areas beyond. The resulting lack of distinctive variation between one place and another is attributable to a number of factors that need to be examined.

The Landscape Has Become Fragmented. Before their explosive growth, the visual edge between town and country, which was the consequence of productive, economic, and functional connections, was clear and well defined. The town drew its character from its regional setting. From within looking out, or from outside looking in, one had no question about one's whereabouts. This sense of context is still apparent today in old towns and cities that have remained unaffected by contemporary economic pressures for growth and change. Examples can be found in old towns in southern France and northern Italy whose economy has stagnated, leaving only an aging population behind. In areas of high growth, however, that condition is reversed. Urban expansion now engulfs a rural environment that no longer serves the productive and symbiotic relationship it once had with a discrete urban place. What is left of the landscape is now fragmented *within* the city, in remnant ravines and river valleys, copses, ponds, and pockets of farmland. Where the land has retained its basic topographic or biological character, either fortuitously or by design, it can maintain something of its original identity. But generally the conditions that created the essential identity of the urban environment have been lost, smothered by suburban expansion that no longer has functional connections with it except as real estate. In addition, in the absence of the cultural imperatives that have in the past created a distinctive urban fabric—such as the great periods of Renaissance city building—the identity of the contemporary city is largely dependent on the character of its indigenous landscape.

Development Expands Outward. It has long been the fate of the rural landscape at the edge of the city to be the raw material for housing subdivisions, industrial estates, and mobile-home parks. The notion that urban development is the highest and best use for nonurban land is written into the lexicon of every urban planner. The changing scene at the edge and the placelessness that goes along with it has become a battleground between efforts to preserve rural land and the relentless forces of urbanization.

Urban Sprawl Confronts Everyone. The city's edge is one of scattered, haphazard development. Questions arise about what one is seeing here. Is this meant to be an urban or rural environment? Is it the beginning of a definitely urban place in the making, or the beginning of a new kind of countryside? Collectively, the corner gas stations, the hamburger franchises, industrial estates, new university campuses, or used car dumps do not seem to have made up their minds which direction to follow. Development seems to defy any semblance of order or integration in the way that has evolved in downtown areas. As *Time Magazine* has commented, "What are these places then? They are a form of urban organization—or, sometimes, disorganization—so new that demographers have not yet coined an accepted name for them. But outside almost every major American city, one or more counties are developing the characteristics of Du Page or Gwinnett or Fairfax County, Va., across the Potomac from Washington, or Orange County, between Los Angeles and San Diego, or Johnson County, Kans., next to Kansas City. These sprawling, increasingly dense suburbs might be called megacounties."[3] Until 1955, when Walt Disney began to build, Orange County's major preoccupation was cultivating oranges. Today, the county's agriculture has been supplanted with development across the entire wedge of land along the Pacific coast between Los Angeles and San Diego, which supports a population of 2.2 million people.[4] The phenomenon of what Joel Garreau has called the "Edge Cities" has appeared as a new form of existence where more than 50 percent of the American population already works.[5] These are "high rised, semi-autonomous, job laden, road clogged communities of enormous size, springing up on the edges of old urban fabrics where nothing existed ten years ago but residential suburbs or cow pastures."[6]

Recent observations of patterns of urbanization made apparent through night photography of the earth by the Environmental Awareness Center at the University of Wisconsin have shown that urban systems in North America, which include cities and their suburban peripheries, are linking up to form rings of urbanization around significant and valuable agricultural land and natural resources. The result is a somewhat circular network of cities across the United States that "has altered our view of how urban areas should be perceived and managed."[7] By identifying all counties from satellite imagery that are bright at night, the center has discovered the configurations of twenty-three constellations. Estimates from the 1980 census for these counties indicate that 80 percent of the population lives within these constellations.[8] It has long been assumed that urbanization should and can be stopped, particularly when the preservation of farmland is at issue. "Yet the 1970s, when the United States tried to do both, proved that the city's impulse to spread is unstoppable."[9] Thus it is clear that the problem of regional identity must be seen from a different perspective.

Perceptions of Lost Identity

Placelessness has many facets. Physically, it has to do with the underlying planning determinants of the highway system, land subdivision and zoning by-laws, and the imperatives of land economics and development that work within this framework. But from the perspective of the issues being explored in this book there are

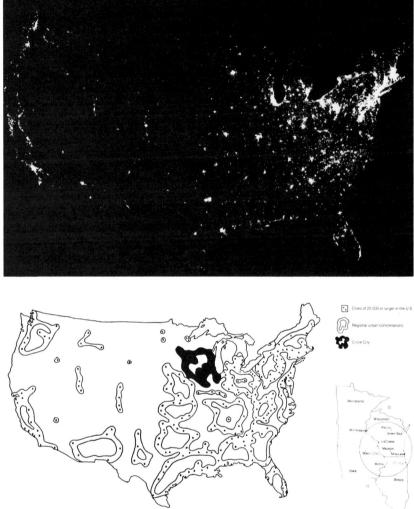

Satellite picture of the United States at night. (Environmental Awareness Center, University of Wisconsin)

Population centers of the United States. Cities of over 20,000 were connected in a manner that made the least impact on identified critical natural resource patterns. (Environmental Awareness Center, University of Wisconsin)

three that should be examined that may provide some insight into the problem.

Unbuilt Space. There are vast areas of open land, some manicured, some simply abandoned, that serve no inte-grative or productive functions. One of the clear indicators of current values toward the urban landscape lies in the way open space is used. Much can be learned from touring the main streets of the suburbs with this in mind. For instance, every newly de-

Is this a new kind of city or another form of countryside? Suburban form is based on zoning and an individual expression of development that lacks connections with its surroundings. The disappearance of street life makes for inhospitable and placeless environments.

signed corporate project manages to perpetuate the facelessness of the suburban environment in several ways.

Each development is isolated on its own plot of land. Its individuality complete, it lacks recognizable connections with its surroundings. Planning guidelines, intended to contribute to the common good, insure discontinuity rather than integration in development, making isolated buildings inevitable. Architecture is confined to the single expression of the building. Street life disappears. As Clay comments, "Each age of confinement takes its toll: traditional

architects and their buildings are so inward-directed, the designers so motivated to produce single, isolated, photogenic, tight-security structures, that street life around buildings becomes a second-class activity. Louis Kahn immortalized buildings by declaring grandly that 'a street wants to become a building'—an architect's self-serving declaration if I ever heard one."[10]

There is little to reflect the continuity of built or natural environments. Under such conditions there can be little recognition of climate. Every tower block and every windy street create hostile winter and sum-

The suburban shopping mall. An isolated social focus separated from other places by the no-man's-land of the highway. All the emphasis is placed on the destination and none on the journey.

mer environments while at the same time a comfortable uniform climate is artificially maintained indoors. The basic tenets of urban design that have helped define the physical environment of the inner city are not the determinants that are shaping the form of the outer city. The enthusiasm among an increasing number of municipalities for annual design awards does little to sponsor identifiable livable urban places because the design criteria that apply are of a different order.

Much of the urban landscape—the parks and gardens and formal open spaces of the city—has been subjected to a universal design standard that denies a sense of place.[11] The ecological diversity of the native landscape is replaced by horticulture. The new plantings that grace the grounds of every subdivision and corporate development are selected from the nursery catalog of best-selling species. They replace the woods and plant communities of the rural environment with exotic alien species, that deny the creation of a context between what is old or indigenous and what is new.

Social and Perceptual Issues. The dynamic of economic and marketing forces that has created the strip and shopping plaza has had social as well as physical consequences. The streets are full of cars but empty of people. The essential nature of the downtown street, with sidewalks, shops, restaurants, theaters and a lively pedestrian environment, is turned upside down in the suburban fringe. Life becomes a series of isolated events, nodes of activity between home, shopping, recreation, and work, made accessible by a no-man's-land environment of highways. For the highway strip, parking is the key element that generates sales and, therefore, must be in full view to make the outlet function. This immediately precludes contact with the street as a social environment. The street becomes a separator rather than an integrator of people, generating a need for another means of social interaction which the sports arena, indoor racetrack, shopping center, restaurant, and pool-hall provide as indoor experience.

Energy and Resources. It has become a tenet of faith that North American cities are places of unlimited and one-way energy systems. It is a belief that gives free rein to attitudes that accept high consumption, pollution, and complete freedom of choice as a normal way of life. The early suburb depended on the streetcar—a fairly energy efficient form of transportation—which was linked to the city center. Commercial life, therefore, naturally centered on the street. The postwar suburb, however, was based on the private automobile and was, therefore, no longer constrained by a fixed mode of transportation. From this, quite different patterns of

growth evolved. Development was no longer dependent on the city; it could spread randomly in any direction and without focus, with cheap energy, cheap land, and prosperity creating the conditions for rapid growth.[12] The suburban landscape, up to the energy and economic crises of the 1970s, increasingly reflected the imperatives of commodity production, and like any other mass-produced commodity, it assumed everywhere a universal image.[13]

Vicarious Experience in Funland

The shopping mall developed from the need to provide a commercial and social focus no longer found in the street system. Since its debut in the Southdale shopping center near Minneapolis in the 1960s, the mall has replaced the old corner store and Main Street as the core of the community—a necessary response to the isolation and loneliness created by low-density development (or sprawl) and the need for personal contact.

Today, all over the world, from the United States and Europe to Singapore and Hong Kong, the mall has become a complete commercial experience. An irresistible magnet, it lures old and young, teenagers and families to spend their money and a large proportion of their leisure time within its covered and air-conditioned spaces. Spreading like

The West Edmonton Mall, shopping cum entertainment center: the vicarious experience in fantasyland. Although immensely popular as an attraction, it lacks any connection with *its location, or with regional environmental influences of climate, natural history, and local culture. (Moorhead Fleming Corban)*

an octopus over vast areas, it includes parking, shopping, and entertainment—and a complete weekend's fun for everyone. As one ardent admirer proclaimed, "twentieth century culture is best epitomized by the mall."[14] It represents contemporary values; consumption as a leisure-time activity, a world where buying, living, and entertainment are indistinguishable. Among the many features of the giant West Edmonton Mall in suburban Edmonton, Alberta, are flamingos and ibises, sharks, Siberian tigers, fountains modeled after Versailles, a full-scale amusement park, a 10-acre (4-hectare) water park with

5-foot (1.5-meter) waves and water skiing, a rising and setting sun for sunbathing, submarines, and a Spanish galleon. This is twentieth-century desire made large, an endless summer in a city where it is cold from the end of September to the beginning of June and where the temperature drops as low as minus 45 degrees centigrade. As the public relations officer proclaimed with great pride to a gathered assembly of expectant visitors, "it can be hot out there, or freezing cold; but in the West Edmonton Mall its always a comfortable, balmy 72 degrees [Fahrenheit]." Now the mall has mi-

The transfer of place and experience from tropical regions to cold ones is a prescription for no-place. (Moorhead Fleming Corban)

grated to the city centers themselves. As a symbol of contemporary urban life it has many implications, several of which are important for the regional imperative.

FOR ALL their make-believe quality, the new megacommercial entertainment centers do express a need for cultural and social opportunities. People flock to them, not only for the shopping and the fun, but also to bank, worship, and receive counsel. Many elderly or infirm people use them for their daily exercise. Although it is clear that the issue of identity is connected to the centers themselves, their impact on the larger environment is perhaps, more crucial. The malls have become a series of unconnected events, islands of activity and life separated by roads, parking lots, and sprawl. Lacking are the pedestrian connections and the social contact that these connections bring. The hostile parking environment surrounding the mall complex serves to accentuate its isolation from the larger environment. It's wonderful once you get inside but pure hell getting there.

The spread of the mall across the world is more than sophisticated retailing or high-pressure advertising.

Inherent in the entertainment it offers to lure people to spend is the vicarious experience of the natural and cultural environment. It tells of a nostalgic past, a fantasy world created from plastic and conditioned air, of interior spaces in cold climates trying their best to appear as exterior warm ones. The caged flamingo, tropical plants and fish, the sun, waves, and beach all simulate tropical environments transported to alien northern ones—a make-believe world that has no connections with the cultural and ecological realities of the place. The transfer of experiences from their places of origin to where they don't belong has become a universal phenomenon of contemporary urban life and a major contributor to the sense of placelessness that massive urbanization has helped to create.

When transferred to alien climates, the perpetual sunshine and temperatures that never change have had a profound effect on styles of living and environmental perceptions. Urban life has become a series of isolated worlds of experience that deny sensory contact with the variables of the environment. It is never too hot or too cold, the danger of the roller-coaster ride is simulated, the wild animals are safe behind bars. It is enjoyment without risk; a Brave New World of real life Feelies; the perfect environment supported by unlimited energy resources and technology in the service of the vicarious experience.

It is tempting but too easy, however, to dismiss these places as terrible examples of how low humanity has sunk in its pursuit of pleasure and escape. Too many people seem to enjoy them. Even those who disapprove often find themselves drawn into the web of excitement that such places exude. "I went there to criticize but ended up by having a very good time," said a friend of mine, hardly wanting to admit to such perverse feelings about the entertainment center he had visited. So the task is to try to understand the opportunities the mall concept presents for learning more relevant environmental and social values. There is little doubt, in fact, that unless the educational experience is fun it is likely to fail in its purpose. The following case study is one attempt to give expression to this idea.[15]

The Shopping Mall: An Alternative to Make-Believe

In a new community of about 50,000 people on the island of Hong Kong a major shopping center was proposed in the early 1980s as an extension to an existing retail and office complex. It included about 100,000 cubic meters of retail space on eight levels enclosed by a glass-roofed atrium, two office towers, and underground parking. Since the shopping-center concept was, at the time, relatively new to Hong Kong, the project posed some interesting questions about the changing social patterns and values of the Hong Kong Chinese, the problems of life outdoors where cramped open spaces provide little relief from the hot humid summer, the nature and function of interior malls in such a cultural milieu, how people use them, and what programs would be appropriate to the larger social and physical environment of the city.

Mild winters and extremely hot, humid summers are typical of Hong Kong's climate. The exterior podiums that have become a standard way of linking apartment towers are little more than inhumane expanses of concrete. As public spaces they lack the most elemental connections to green plants, and as environments they express acute sensory undernourishment. In contrast, every highrise apartment balcony is crammed with anything that will produce a green leaf, or a flower—a sure indication that the Chinese yearn for living things to offset their extremely crowded conditions. The air-conditioned interior mall provides a pleasant, drier indoor climate and is a marked relief from such places. Like its counterpart in the West, the mall has become the answer to greater, more comfortable shopping for people on foot. Chinese families spend their one nonworking day of the week in these air-conditioned places, shopping, eating, socializing, and keeping cool.

It was apparent that developing a program for what was in effect a day's outing for a family, was more than simply providing a place in which to spend money. The objective was to provide a gathering place and leisure center, for large families with children, for teenagers and young adults on their own. In addition, it was important to imbue these varied experiences with a sense of purpose, where fun and variety could be integrated with the learning and sensory enjoyment that is missed in the cramped urban conditions of Hong Kong. Plants and flowering trees, small-scale agriculture, quiet places, sensory rich places for play were important as were opportunities to learn useful skills and crafts and to have exposure to often forgotten historic traditions.

The concept for the complex evolved from these social and environmental considerations. It was based on the notion of creating within the complex a minicity, with parks and open spaces, public and private areas, and varied environments that would cater to the recreational, entertainment, and educational needs of everyone. The following elements shaped the design.

The main atrium space was conceived as a social center for meetings, entertainment, fashion shows, dining, and destination points, within a landscape of water, plants, and beautiful paving. Workshops where computer technologies, calligraphy,

A proposed shopping center for Hong Kong.
(Dimitri Dimakopoulos and Partners: Hough Stansbury Woodland).

*Design and pro-
gramming of its
interior spaces were
aimed at enter-
tainment that was
relevant to the
local scene and its
social priorities.*

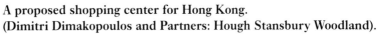

*The atrium, de-
signed for shows,
entertainment, and
meeting.*

*Scent gardens and
flower shows at
different seasons
are important to
the Chinese in a
crowded city.*

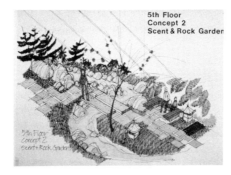

*Workshops for
young people to
learn traditional
and modern
crafts—pottery,
calligraphy, com-
puter technology,
painting.*

pottery, painting, and various trades could be learned and practiced were included for teenagers and young adults. A fifth-floor garden included a minifarm, water and flower-display gardens, a series of playgrounds and small parks for small children and parents. There were exhibits and demonstrations of the latest appropriate technologies in energy production and efficiency and do-it-yourself kits. A shaded outdoor sculpture garden for seating and revolving art shows served as a focus for meeting friends in a pleasant and cool environment shaded by a canopy of trees.

As an alternative to the often meaningless, vicarious experience of so many entertainment-cum-shopping malls, this project was intended to make connections between people and the larger urban environment in ways that are relevant, useful and part of the place, while at the same time providing an educational experience that is based on fun and enjoyment.

The Imposition of Social Values

Apart from questions of planning and environment, social factors have always affected issues of identity. The migration of middle-class whites to North American suburbs in the 1950s brought with it a requirement to conform to certain standards of behavior and aesthetic proprieties that usually had nothing to do with

practical needs of privacy, personal life, or convenience. The location of the house on the lot, fence heights separating backyards, tree planting, and the front yard all had to conform to the new subdivision's design requirements of unity and harmony. Some of these requirements were built into township by-laws, and many others were the result of community expectations and aesthetic conventions. The imposition of conformity by the few on the many leads to a denial of social and environmental diversity—those qualities that make ethnic downtown neighborhoods so enjoyable and interesting.

Some years ago a friend of mine moved into a new residential neighborhood on the outskirts of Winnipeg. Determined to create an outdoor environment that suited his own needs and tastes rather than those of his neighbors, he decided to return his front yard to a landscape of native prairie woodland rather than to one of turf.[16] He planted Manitoba maple, Ash, Choke cherry, Prairie rose, and native grasses gathered from the countryside, and over five years the little woodland established itself and began to look very beautiful. For him, it was a poignant reminder in miniature of the native landscape that he loved and that had been there before.

But he had problems with his neighbors; his front yard stood out so obviously from the others on the street. It expressed not only the native prairie landscape but also the personal tastes and inclinations of an individual. He recalls his first altercation with his immediate neighbor. "During that first summer [he] paid me a visit and asked when I intended to cut my lawn. Not really knowing much about the gentleman and not taking his statement seriously, I retorted that lawn cutting was a waste of energy and I was planning on getting a goat. His jaw dropped . . . and once he had regained his composure he stormed off muttering something under his breath. A few minutes later I looked out of my window and noticed a lawn mower sitting in the middle of my front yard. Thus was my neighbor's subtle response. Needless to say we did not get along very well from that point on."[17]

Some years later my friend and his family moved to a larger house. He returned a few months later for a visit and found to his astonishment and chagrin that the new owner had bulldozed the entire prairie landscape and replaced it with a nice lawn—just like the others on the street. This experience is a poignant demonstration of just how firmly entrenched conventions can be and how influential they are in maintaining a homogeneous cultural and physical landscape. They are the constraints of imposed aesthetics for the common good. Time, however, is the ingredient that softens the facelessness of the new in both a social and a physical sense. As Barrie Greenbie points out,

"The often deplored 'look-alike houses' of North American suburbs are as much a function of the physical and social mobility of the modern middle class as of mass taste and mass production. Houses that are really lived in for long do not look alike, any more than do the people who live in them."[18]

The Transient Society

Alvin Toffler has said that transience is the basis for the fast-forming society of today, in where people both work and live.[19] Over a lifetime, the average American will have traveled 3 million miles (4.8 million kilometers) or more, more than thirty times the total lifetime travel of his 1914 forebears.[20] People in the developed countries now have access to public services, leisure, travel, education, instant communication, modern living standards, plentiful and varied food they do not grow themselves, all made possible by advances in technology and abundant energy. The wide variety of options available to everyone in democratic societies is in marked contrast to their absence in the preindustrial era. "When benefits were limited to the few their provision had an equally limited effect on the landscape."[21] But now the impact of those benefits on the landscape as a whole is proportionally larger and more devastating. Adherence therefore, to the old regional imperatives that created the uniqueness and sense of place of past landscapes has now lost much of its original purpose. Whether the sense of place is important to contemporary life becomes a question of societal values rather than necessity. Yet the consequences of today's values on the environment are only now being recognized.

FOR TODAY'S traveler having breakfast in London, lunch in Rome, and supper in Majorca is no longer the joke it used to be. Nonetheless, the mythical notion of the American tourist leaping across continents to savor the traditional delights of local cooking and culture, complete with bermuda shorts, camera, and sunglasses, or the more modest Britisher with his jam sandwiches and fieldglasses in search of the local bird sanctuary, is, of course, a stereotype that has had its day. It does, however, express a universal urge to go somewhere different, to live in one place and work in another, that travel now makes possible. This is in marked contrast to other less wealthy and place-bound countries such as China, where the average citizen's first job fixes his place of residence for the rest of his life.[22]

Because of this desire for mobility, a large part of our industrial economy is predicated on the need for transportation—automobiles and highways, airplanes, public transit systems. The ultimate goal of all this technology is speed and convenience. Gone are the days when one traveled the seas by "steamer," as my generation did not so long ago, taking eight to ten days to cross the Atlantic, depending on route and weather. The trip today is cheaper and faster, and takes only eight hours by air.

The industrial revolution began a process of change to the landscape that has fundamentally altered our perceptions of nature and regional differences. The railway, the automobile, and the airplane all structure the experience of the environment and separate us physically and in time from the world through which we pass. Urban expressways and regional highways have had enormous physical and social impact on the places where people live and on the sense of identity and connection to those places. Old ideas and values about what is important about a place, which evolved from the limitations of getting around, have fundamentally changed. The understanding of places increasingly becomes a matter of specific experiences—the airport one leaves and the airport where one arrives—with no link between experiences.[23] Travel permits a new perception, a world view of the environment that crosses regional boundaries. And so for the first time we have the opportunity for a new and large-scale sense of regional diversity. But the process of revealing it also tends to destroy it.

The Travel Corridor as Placeless Landscape

The degree to which our experience of the environment is dominated by the travel corridor is often not appreciated. Frank Egler observes, in fact, that there is no land we see more often.[24] In the United States the right-of-way domain—that strip of land devoted to the movement of vehicles, trains, pathways, communications, and power—has been estimated to cover some 50 million acres of the forty-eight contiguous states.[25] Whether the environment is urban, wilderness, farmland, prairie, or mountains, the traveled right-of-way is how we see the world most of the time. It is here that one would expect regional differences to be readily understood and sharply focused, since the speed of automobile travel reduces the physical scale of the landscape. The opposite, however, is more often the case. Highway alignment ignores topography and natural features, and the design of roadside planting is based on predetermined, universally applied lists of plants that have no relationship to the surrounding vegetation. Vegetation management, based on the use of herbicides to control "brush," and mowing to turf to control "weeds," creates uniform treatment that negates the inherent regional differences between one landscape and another. The emergence of herbicides, in fact, coincided with the 1956 inauguration of the 70,000 kilometer interstate highway system in the United States.[26] With some exceptions, the design of the highway landscape is truly international in its style, reinforcing the isolation from the outdoors that the interior, protected environment of the vehicle itself initiates. Like many other rights-of-way it contributes nothing to the health or diversity of the environment through which it passes.

Some efforts to integrate the highway into the regional landscape have been made. In Germany, for instance, the character of the Ulm-Baden-Baden autobahn moves through undulating fields of corn that alternate with woodlands: "The road, even when seen in the full width of its double carriageways and center-strip, is in scale with the general pattern [of the countryside]."[27] The Blue Ridge Parkway, in the section of the Appalachian Mountains in Virginia, incorporated controlled access, broad medians in a wide right-of-way, the conservation of native vegetation, and the careful alignment and placement of the roadways to insure a fit with the terrain. Similarly, the Taconic State Parkway extending north from New York City is internationally known as an example of landscape integration where the four-lane parkway was conceived as two separately aligned highways.

For the most part, though, highways are insensitive to the character of the landscape, cutting across its grain in their quest for the straightest route. The properties that make the country lane so much a part of its environment have to do with alignment (it follows contours and avoids obstacles rather than cutting through them) and scale (it is closer to the surrounding landscape). As Tunnard comments, "It is the immediate proximity to nature that makes driving on small country roads so delightful: the feeling of foliage overhead, of grass next to the wheels, of fields, hedges, and trees almost within reach, . . . the fact that the roadside occupies as much as 80 percent of one's field of vision."[28] The country road is largely the consequence of necessity, of doing things the easiest way. However, modern highway engineering often reveals the geological character of a region by default rather than by design, exposing rock formations through deep cuts in the land. For instance, along the northward route from the limestone regions of southern Ontario to the granite formations of the Precambrian Shield, the first significant indicator that the boundary between the two has been crossed is a massive granite rock cut some twenty meters high through which the road passes. It is an unequivocal signal to the driver that he is now in Precambrian Shield country, even though the actual boundary is some ten to fifteen kilometers south of that point. Vegetation, controlled more by climate than by geology, is slow to change and less dramatic as an in-

dicator. Similarly, the highways in the state of Georgia often reveal the characteristic red soil of the region through erosion of the road verges.

In the 1920s and 1930s, an aesthetic desire to improve "ugly roads" led to beautification programs in North America and to the creation of the "Landscaped Parkways." These were lavishly planted showpieces of landscape design based on the horticultural tradition, of which the Merritt Parkway in Connecticut is an early example. The consequence of this international style of highway landscape was the spread of ornamental trees and shrubs across entire regions on a vast scale that denied rather than expressed the character of the surrounding landscape. The urban parkways in Ottawa that were part of the 1950 National Capital plan continue that tradition of horticultural beautification, with ornamental plants dotted around enormous expanses of exquisitely maintained turf. As greenways and links to the city center from the surrounding region they do their best to smother the extraordinary drama of Ottawa's stratified limestone rock formations.

The phenomenon of the highway strip, the familiar commercial development that can be found at the outskirts of every town, has created another visual no man's land. As a contemporary landscape, it is the product of highway travel, an extension of urban economic life and market forces with roots that go back, in

A shale cut in Kentucky. Highway engineering ignores topography and natural features but fortuitously establishes strong connections with the surrounding landscape by revealing the underlying geology of the region. Note the emerging juniper plant succession, another indicator of regional identity.

Granite rock exposed in an Ontario road cut establishes the fact that the driver has entered the Precambrian shield.

North America, to Indian paths, wagons, and the internal combustion engine in the 1920s.[29] In terms of our understanding of landscape patterns, the highway strip has urban parallels with the landscaped rural highway. It defies regional identity, but with the important distinction that the rural highway's placelessness is the consequence of conscious design and management of its right-of-way vegetation, whereas the strip is today's new vernacular. But as Clay has remarked, "the strip is trying to tell us something about ourselves; namely, that most Americans prefer convenience; are determined to simplify as much of the mechanical, service and distribution side of life as possible . . . The value systems of the strip derive from the open road rather than the closed city. [They] become the city's vital contact zones with surrounding regions and their fast changes reflect population and taste shifts on the urban front."[30] The strip lacks identity less by what it is than by its endlessness, its absence of contrast and reference to the landscape. Disconnected and placeless as the strip appears to be from an environmental perspective, it is nevertheless the symbol of the choices that become available when affluence, cheap energy, and shifting ways of life are the motivating forces of an urban society. It exists because it is convenient and practical to today's needs—motivations that also underlay the evolution of earlier vernacular landscapes.

The appreciation of the environment around the strip and the clues that tell us something about a place's differences are relegated to the museum rather than the physical environment; to formal education rather than knowledge gained from direct involvement with nature. Two experiences I once had in Florida brought this fact abruptly to my attention. The first was a visit to the Natural History Museum at the University of Florida at Gainesville, where an excellent exhibit of four major Florida ecosystems was on display. They included the Coral Reefs, Mangrove, Longleaf pine / turkey oak, and Mesic Hammock systems. The exhibits demonstrated with lifelike mock-ups and clear, well-written texts the interrelationships between life forms and their significance to Florida's natural landscape. While I was there I saw several groups of schoolchildren being led through the show. They were listening attentively to enthusiastic explanations by their teachers of the complex interactions and relationships of these ecosystems. As the exhibit explained at the entrance, "Science, like many other beliefs, is an adaptive form of human behavior that attempts to interpret and order our environment."

My second experience followed shortly thereafter on a drive down Highway 441 to Orlando in search of these natural marvels. Almost dead straight and with little topographic change, the road passed first through

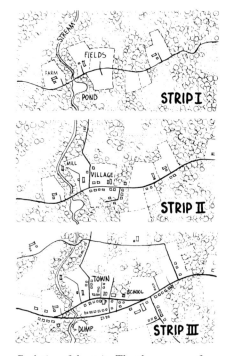

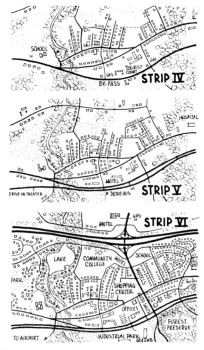

Evolution of the strip. The phenomenon of the highway strip is a reflection of contemporary values. Its placelessness, however, is more often a function of endlessness than of form and function.

Experienced in relationship to contrast, gateway, and links with the native landscape, the strip can be understood as contemporary vernacular. (From Grady Clay)

open farmland and evergreen woods, its blandly landscaped turf, wide median, and verges oblivious to the changing scenery. The road then entered a continuous strip development of motels, restaurants, and car lots, all advertised by a confusion of billboard displays that dominated my cone of vision. It was difficult to decide whether I was approaching or leaving a settlement since the density and scenery never changed. Apart from the flatness of the landscape

and sporadic tropical vegetation that might give me a clue about the identity of Florida, the trip told me nothing of the places and diverse environments I had previously learned about in the museum at the university. It did, however, tell me a great deal about Florida's booming tourist industry that accounts for some $15 billion a year, but at the same time denies the tourist the opportunity to experience the very environment he or she has come to see. But then, in

A strip in Mississauga, Ontario.

central Florida the tourist is looking for the Epcot Center and Disney World, not for nature. And among the delights of Disney World he or she will indeed find "nature" in a vastly enlarged plastic extension of the museum exhibits at Gainesville.

These two experiences taught me a lot about current values. They also highlighted our schizophrenic attitudes toward economic growth and prosperity and the preservation of natural heritage. Connections between these two notions are rarely made. To understand the latter has become a question of the vicariously experienced indoor exhibit, remote from the place, formally taught, with simulated plants and animals behind glass screens in air-conditioned comfort. The highway corridor tells us nothing about those living systems or about how they look and feel as unique environments. It expresses prosperity and the isolated experience at each end. The travel corridor today reflects current priorities that must be addressed as a major issue for environmental design. Careful observation also reveals that there are aspects of this landscape that can provide a basis for action.

Alternative Design Strategies

Since our perceptions about the environment are largely formed by the routes we take from place to place, the travel corridor becomes a way of making the traveler aware of the changing environments around him. It is also a potential form of education, as Appleyard has suggested, where the landmarks, history, natural and cultural shape, and evolutionary processes of the world are revealed.[31] While the highway right-of-way has become a dominant force in reinforcing placelessness, it is also potentially the most powerful agent for reinforcing the local and regional character of the surrounding landscape.

An essential aspect of this potential lies in the notions of contrast, change, and gateway between one environment and another. Key points of identity along the travel corridor, where the experience of one place gives way to the next, are necessary to how we identify our whereabouts. In the urban region, these might be found at highway interchanges, bridges over river valleys, or sections of the highway that link the airport to downtown. Whether they are classified as "beautiful" or "ugly," they generate interest and understanding by echoing the variety of the landscape through which the corridor passes. These expressions of identity are best revealed when the management of right-of-way vegetation reflects the uses and biological characteristics of the land. For example, some prairie provincial highway departments enter into agreements with farmers to mow the grass within the right-of-way. This practice provides economic benefits to both the farmer and the department, and it extends the working landscape of the region into the road right-of-way itself. It helps create a sense of harmony or fit between the two. "The aesthetic experience is merely a benign end result of a quite different approach to the freeway journey which builds upon content and process in the landscape."[32] The biological value of the right-of-way can also be seen as a potential vegetation corridor for wildlife, linking remnant natural habitats, such as woodland, that have been isolated by agricultural development. This approach is being widely adopted in Britain where motorway rights-of-way are planted and left to regenerate naturally. These refuges have, in fact, become important new linear habitats for protecting wildlife in a farming countryside that is rapidly losing its hedgerows.

Two case studies, both undertaken in 1987, have explored the potential for creating a sense of identity in the highway environment. The first, a study of a rural highway in Ontario, explored the application of environmental management guidelines to corridor vegetation along a twenty-one kilometer section.[33] The study area was a rectilinear settlement pattern of agricultural fields superimposed

Toronto's 401 expressway passing over the Don Valley creates a sense of identity at the interchange.

A strip in Eugene, Oregon. A sense of visual identity is created with the forested hills that dominate the landscape.

A gateway in Windsor, Ontario, announces the entry into one part of town from another.

over a naturally rolling terrain. The highway followed this man-made pattern, ignoring the terrain and maintaining an unerring straight alignment throughout its entire length. Its cross-sectional design also left wide verges on either side. These had been allowed to naturalize as woody growth where a woodland seed source was available, and to herbaceous growth near agricultural fields. Agricultural land abutting the road showed considerable variation in cropping patterns, which included corn, grain, hay, and grazing. Forest cover and wooded stream valleys occurring in part of the study area formed links to other natural systems. Analysis of the potential for alternative management of the right-of-way included:

Areas adjacent to open field hay systems could be managed for hay mowing depending on the slope conditions of the right-of-way.

Natural regeneration could be encouraged where the right-of-way made potential links with existing woodland and ravine systems.

Natural regeneration made sense on steep cut and fill slopes where hay mowing was impractical, and on abandoned agricultural fields.

Natural regeneration and reforestation within the loops of highway interchanges could reduce maintenance costs and create points of identity at decision points. Hay mowing could be pursued in the larger areas outside the loops.

Modification to the high maintenance of service centers to naturalization, haying, and forestry would have great potential for interpreting alternative highway management approaches and of surrounding land use, where "redirected" rather than "reduced" maintenance could be explained to the public.

The second study was focused on an urban setting. The Ottawa River Parkway is an important scenic drive into the center of Canada's national capital and follows one of the greenway corridors laid out in the 1950s as part of the city's plan. Conceived in the gardenesque tradition of the early American parkways, it was planted to cultivated trees, shrubs, spring bulb displays, and acres of sweeping lawns. Because of government cutbacks, an investigation of cost-saving alternatives to this horticultural landscape was undertaken in 1987 and resulted in an alternative approach to urban vegetation management.[34]

The study addressed several critical issues in addition to the question of effecting economies in the maintenance of this high-cost landscape. Among these was the issue of diversity. The uniform horticultural development of the national capital's 2700 hectares of parkway landscape at present provides little variety from one place to another. Equally important was the need to create a sense of place. One of the great experiences of Ottawa for the visitor to the city is the splendid natural setting that frames its monuments, public buildings, and drives along the river. Where the landscape has been left alone, the stratified and exposed limestone rock, the rapids, falls, and fast flow of the river, its marshy edges, and native woodland provide dramatic and unparelled views of the city and its skyline as they unfold along the route. This landscape is what makes the city memorable as a beautiful place. Yet in spite of this extraordinary

Much of the natural landscape that gave character and identity to Ottawa had been erased by years of horticultural maintenance.

The limestone formations provide a powerful and dramatic entrance to the city.

A vegetation study for the Ottawa River Parkway. An underlying purpose was to re-establish the drama of the river landscape by restoring the native landscape. (Hough, Stansbury + Woodland)

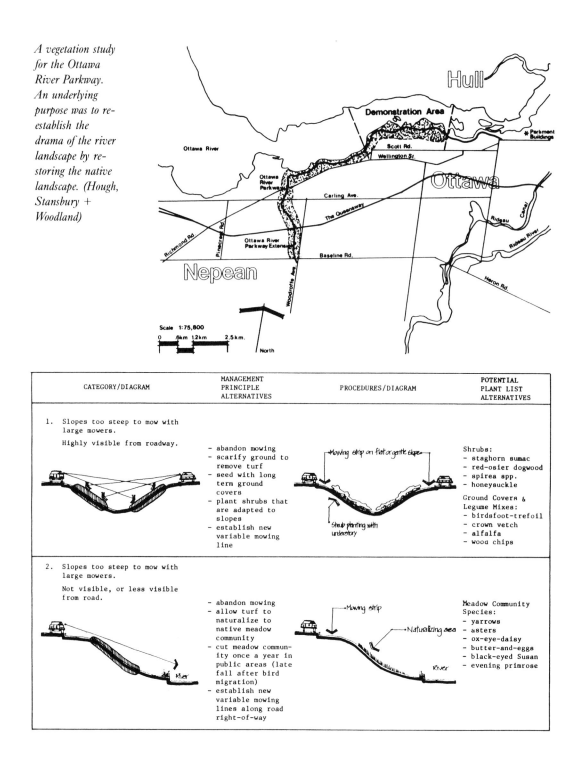

CATEGORY/DIAGRAM	MANAGEMENT PRINCIPLE ALTERNATIVES	PROCEDURES/DIAGRAM	POTENTIAL PLANT LIST ALTERNATIVES
1. Slopes too steep to mow with large mowers. Highly visible from roadway.	- abandon mowing - scarify ground to remove turf - seed with long term ground covers - plant shrubs that are adapted to slopes - establish new variable mowing line	*Mowing strip on flat or gentle slope* *Shrub planting with understory*	Shrubs: - staghorn sumac - red-osier dogwood - spirea spp. - honeysuckle Ground Covers & Legume Mixes: - birdsfoot-trefoil - crown vetch - alfalfa - wood chips
2. Slopes too steep to mow with large mowers. Not visible, or less visible from road.	- abandon mowing - allow turf to naturalize to native meadow community - cut meadow community once a year in public areas (late fall after bird migration) - establish new variable mowing lines along road right-of-way	*Mowing strip* *Naturalizing area* *River*	Meadow Community Species: - yarrows - asters - ox-eye-daisy - butter-and-eggs - black-eyed Susan - evening primrose

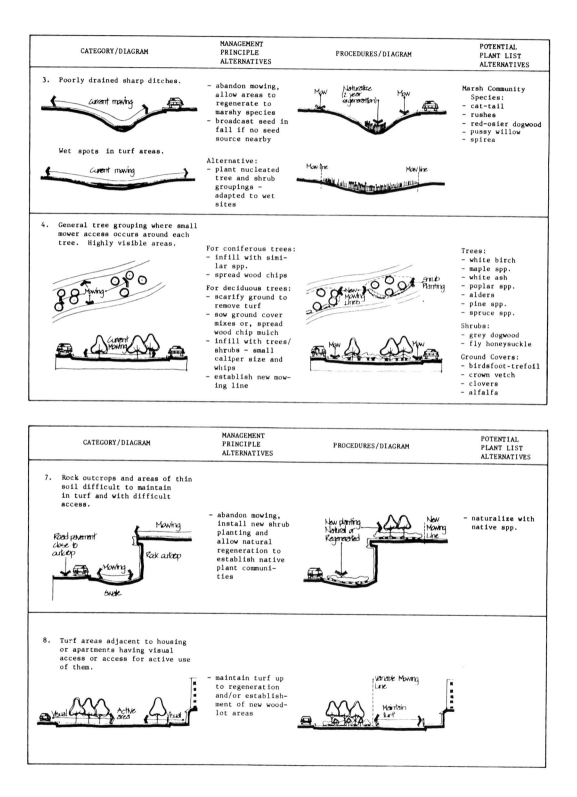

CATEGORY/DIAGRAM	MANAGEMENT PRINCIPLE ALTERNATIVES	PROCEDURES/DIAGRAM	POTENTIAL PLANT LIST ALTERNATIVES
3. Poorly drained sharp ditches. *Current mowing* Wet spots in turf areas. *Current mowing*	- abandon mowing, allow areas to regenerate to marshy species - broadcast seed in fall if no seed source nearby Alternative: - plant nucleated tree and shrub groupings - adapted to wet sites	*Mow Naturalize (2 year regeneration) Mow* *Mow line Mow line*	Marsh Community Species: - cat-tail - rushes - red-osier dogwood - pussy willow - spirea
4. General tree grouping where small mower access occurs around each tree. Highly visible areas. *Mowing* *Current Mowing*	For coniferous trees: - infill with similar spp. - spread wood chips For deciduous trees: - scarify ground to remove turf - sow ground cover mixes or, spread wood chip mulch - infill with trees/shrubs - small caliper size and whips - establish new mowing line	*New Mowing Lines Shrub Planting* *Mow Mow*	Trees: - white birch - maple spp. - white ash - poplar spp. - alders - pine spp. - spruce spp. Shrubs: - grey dogwood - fly honeysuckle Ground Covers: - birdsfoot-trefoil - crown vetch - clovers - alfalfa

CATEGORY/DIAGRAM	MANAGEMENT PRINCIPLE ALTERNATIVES	PROCEDURES/DIAGRAM	POTENTIAL PLANT LIST ALTERNATIVES
7. Rock outcrops and areas of thin soil difficult to maintain in turf and with difficult access. *Road pavement close to outcrop Mowing Rock outcrop Mowing Swale*	- abandon mowing, install new shrub planting and allow natural regeneration to establish native plant communities	*New planting Natural or Regenerated New Mowing Line*	- naturalize with native spp.
8. Turf areas adjacent to housing or apartments having visual access or access for active use of them. *Visual Active area Visual*	- maintain turf up to regeneration and/or establishment of new woodlot areas	*Variable Mowing Line Maintain turf*	

Less costly and environmentally sensitive management will, over time, restore the natural drama of this landscape.

Changing Values and Changing Urban Form

natural heritage, it has been said that Ottawa is a city of views but no places. Much of its designed landscape negates rather than enhances this sense of identity. While the value of a more diverse approach to the management of vegetation lies in the necessity for economy, it also leads to a more self-sustaining landscape and enhanced sense of the region.

The principles for management that were developed were simple and pragmatic. They included strategies for mowing (including steep slopes, river banks, natural areas, and places that flood periodically); infilling of widely spaced trees and shrubs to create habitats and reduce maintenance; and management of natural features such as exposed rock. Also included were design strategies for shaping mowing edges, creating seasonal variety, and similar factors to convey the message that the management alternatives emphasize redirection and sense of purpose rather than abandonment.

Like nature's processes cities change and evolve, and it is the underlying nature of this change that provides us with clues for alternative approaches to design. The evolution of the suburb shows us how rapidly change can occur. During the 1970s, with land, energy, and credit becoming scarcer and with the beginning of a shift in values, the era of massive suburban expansion was greatly slowed.[35] In North American cities the trend has been to move away from postwar urban sprawl to groups of interdependent "urban villages," which are business, retail, housing, and entertainment focal points within a low-density cityscape.[36] Many of the established urban villages in Los Angeles are gaining identities of their own. As an example, the entertainment industry, traditionally located in Hollywood, is moving to Universal City/Burbank. Insurance companies, traditionally found in the mid-Wilshire Boulevard district two miles west of downtown have relocated in Pasadena.[37]

There are various reasons for the shift in focus from sprawl to urban village. One is that the shift from a manufacturing to a service and knowledge base has created a demand for office space. People are more willing to live near the office than a dirty, noisy, and visually unattractive factory. Another reason relates to changes in transportation patterns that favor trucking over rail for the shipment of goods, and automobile commuting over mass transit. A third has been advances in telecommunications that permit more and more work to be done by cheap, long-distance telephone, overnight mail, computer modems, and telecopiers.[38] There are also signs in the late 1980s that the private car is no longer the only form of transportation dictating built form. There has been a movement back to the cities by many middle-class suburbanites. At the same time, ethnic groups are migrating to the once middle-class suburbs, which are now showing signs of increasing their density. Recent trends in planned development in the United States indicate an effort to focus on the street as a pedestrian environment, a return to the street grid, and the establishment of environments over which people have some control: in effect, a return to the traditional ways in which communities have evolved.[39]

EVERY CITY is a mosaic of environments reflecting the complex interaction of natural and social forces which provide the basis for natural and cultural diversity. The city and suburb are two different places and operate on different principles. Identity in the urban center is based on the continuity of the built environment—a matrix of built form. Urban spaces, squares, parks, streets, and

the ways these are linked are the organizing framework. The life and activity of this fabric is nourished at its edges by shops, cafes, cultural and commercial activities. In the suburban and outlying areas of the city, however, the situation is generally reversed. The matrix becomes unbuilt space within which there are buildings. Richard Hedman has suggested an injection of urbanity at key points as a possible method of creating a more decisive impression of order.[40] Planning guidelines for some older suburban centers are encouraging a transition from the typical patterns of the commercial strip—parking and billboards separating retail outlets from the highway—to shops that directly face the street as a means of bringing pedestrian life back to the street environment.[41]

In the suburban landscape as a whole, however, there is a greater potential for establishing coherence and order in the continuity of unbuilt space. Urban design has new parameters where the landscape itself is the dominant element. Identity is established by the density and continuity of its canopy vegetation, by existing and newly created woodland and natural areas, by productive allotment gardens, and recreation spaces. The organizing framework of this landscape matrix lies in a system of connections that include river valleys and streams, hedgerows, and other agricultural land patterns, regenerating and disused rail corridors,

**Some options
for suburban
landscapes:
Toronto.**

*A canopy of vege-
tation provides a
matrix within
which there are
buildings.
(Deborah McNeil)*

and utility rights-of-way. Where these can be exploited as a basis for a landscape conservation strategy that precedes urban growth and gives it form, much of the placelessness of new development could be avoided. Where urbanization already exists, the necessity of establishing a landscape structure with the resources available, of restoring degraded land to a state of health, becomes all the more critical to the creation of sensory enrichment, delight, and sense of place.

Bringing together rural and urban values in the management of urban places is necessitated by what is already occurring: a blurring of the distinction between established ideas of what is urban and nonurban and a tendency toward biologically sustainable land management. Parks are an integral part of this trend. Traditionally, the planning of urban parks systems has been based on a hierar-

chical classification that describes their sizes and uses from the corner parkette to the regional park.[42] This approach belongs more to the realm of orderly planning on paper than to the realities of actual places, or to the great majority of open spaces not classified as parks, but which often perform vital environmental and social functions. Inherent too in the perception of city space is the assumption of equal access everywhere. In these two fundamental planning doctrines lie the basic elements of the placeless urban landscape. They ignore a basic tenet of the regional imperative—that differences in places are the consequence of the biophysical and social forces acting on them. What is needed instead is a response to the city's environment that recognizes its inherent diversity rather than negating it with outdated doctrines: a classification system based on natural and cultural process that

An emphasis on regenerated woodland and small wetland areas creates a continuity of open spaces and corridors.

Some parklands are being naturalized with meadowland as in North York's parks in metropolitan Toronto. (Bill Granger)

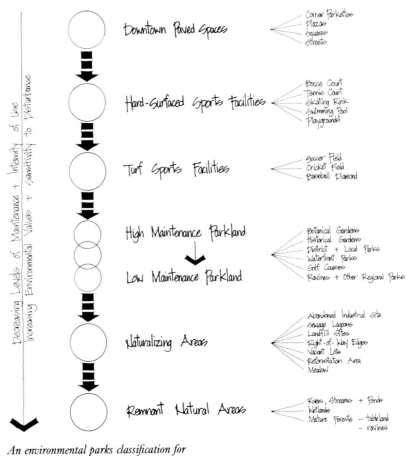

An environmental parks classification for
cities provides the basis for great natural and
urban diversity. Its categories of open space
range from urban wilderness where natural
values predominate to highly manicured
gardens and urban squares where social
values have priority and large crowds can be
accommodated. (From Hough and Barrett)

Tommy Thomson Park. An urban wilderness on Toronto's waterfront.

Horticultural gardens in downtown Toronto. (Suzanne Barrett)

cuts across the artificial boundaries that separate nature and people. Such a system would have a wide range, accommodating remnant or evolving landscapes sensitive to human presence and requiring preservation and controlled access to intensively used downtown spaces where people and crowds prevail.

As I suggested in *City Form and Natural Process*,[43] the larger regional open spaces are often of greater environmental significance as places for the maintenance of wildlife and plants than they are for recreation. Many such places exist fortuitously in undeveloped river valleys and waste places on the edges of most large cities. The Dutch have set aside urban wilderness areas that exclude the

general public but which are used for preservation and research.[44] At the other end of the spectrum the higher-density urban parks are where cultural activities assume priority. Biological classification recognizes recreational, productive, and environmentally significant landscapes in ways that are similar to established classification programs for national parks and biologically significant areas. It comes to grips with the reality that cities have become integrated with their regions and can no longer be seen as separate entities.

On another scale there are opportunities for acknowledging the uniqueness and variety of the city landscape and the diversity of its users. There is growing interest in creating new forms for public and private space where forest, meadows, and wetlands become the key identifying elements of the city's place in nature. Until it was disbanded in 1987, the Greater London Council promoted environmental awareness in Britain by establishing ecological reserves throughout the city.[45] A wave of national enthusiasm for wild places swept the country with the growth of organizations such as the Urban Wildlife Group, as well as television programs and books on how to create wildlife habitats in urban areas. The notion that the cities today are environmentally healthier places and potentially more diverse than the countryside has long been recognized by the Dutch, the Swedes, the Swiss, and in many other European countries. I remember as a child being caught in the Manchester "pea soupers"—fogs that were caused by industrial air pollution and uncontrolled burning of coal in domestic fireplaces. This condition has now dramatically changed in many countries with the postwar introduction of air and water pollution controls. As Chris Baines has commented, speaking of British cities, "our intensive farming countryside is now more polluted with chemicals than much of the green spaces in our cities. Only rarely is anyone spraying the railway embankments with pesticides. No-one is emptying sacks of nitrogenous fertilizers into the canals and, not surprisingly, as towns have been cleaned up, wildlife has benefited."[46]

In Canada, increasing maintenance costs and an interest in promoting ecologically varied urban environments have encouraged some municipalities to enrich their parks. The city of North York in Ontario, for instance, is permitting many of its woodland parks, whose trees were destined for extinction, to regenerate to naturally sustainable forest. Originally degraded with a turf cover and used as if they were standard open parkland, these areas have now been changed to support activities appropriate to a forest landscape. Since 1978, the city of Sudbury, whose native landscape was severely damaged by early ground smelting of nickel ores, has begun the process of reforesting over 1,500 hectares of public and private open space with plants native to its northern region.[47] This city, set in a powerful Precambrian Shield region of granite hills and mixed forest, is once again reestablishing its relationship to its regional landscape.

The identity of the city is a combination of cultural and natural history, the variety of its ethnic and interest groups, its cultural and economic history, and its development patterns. It is, therefore, necessary to see its open spaces in this light, as functional, richly diverse, and environmentally beneficial to the health and productivity of people and places. The task of design is to recognize and build on this inherent diversity wherever the opportunity arises.

6 / Industrial Landscapes and Environmental Perceptions

Landscapes of power. The technologies that support the city have spread into the larger landscape.

The extension of the city's support systems into the countryside is one of the major phenomena affecting the modern landscape. The changes occurring today, such as the clearing of forests in favor of agriculture or the draining of marshes, are different in a number of respects from those transformations that occurred in the past. They differ in the scale and speed with which they happen, in the variety of industrial objects and land uses that are being imposed on the land, and in their wide distribution over the entire region. At the same time, traditional agriculture and for-

estry are no longer sustainable rural occupations contributing to the countryside's health and beauty. They have, like the cities themselves, become high-production technologies that are reshaping the landscape and radically altering our perceptions of its regional identity.

The concentration of industrial processes in one place and their direct connections to urban areas are increasingly less relevant to the new high-tech systems that sustain urbanization. Sylvia Crowe has commented that "Industrial expansion which in the 19th century followed

New forms in old landscapes. Regional power grids and other urban technologies have radically altered former perceptions of what is urban and rural.

the railway and the coalfields, today runs foot-loose along the roads and powerlines."[1] Travel, electronic communications, industry, power generation and distribution, television towers, and other urban-based technologies have become a part of the rural areas. The need for safety and for large quantities of water for cooling has determined the location of nuclear power stations far from urban centers and in the proximity of lakes, rivers, or oceans. Communications relay systems for telephone, television, and radio require heights of land free of interference. Electrical utility lines and gas and oil pipelines—the great tentacles of the new city region extending into the hinterland—travel across hundreds of miles of terrain from their points of origin to their destination points of consumption in the cities. Visual perceptions of the countryside are now totally altered. It is almost inconceivable to think of traveling along a road that is not lined with telephone poles and lines. In the vastness of the prairies, for instance, they form the dominant and, for some, comforting acknowledgment of man's presence.

Nonetheless, images of the past persist, in spite of the transformation of the landscape. Engineering structures—the visible expressions of our urban way of life—are something we spend enormous amounts of time and ingenuity trying to conceal on the basis that they mar the beauty of the landscape. In the developing

*Iqaluit, Baffin Is-
land. Global com-
munications are
now an inherent
part of every land-
scape, no matter
how remote.*

world, however, the landscape of energy technology is likely to be seen as
a blessing rather than an aesthetic liability. Power lines and relay stations
are important cultural symbols to a
country trying to modernize. It is
clear that we have not come to terms
with the new landscapes that are
being created. New forms overpower
an old, small-scale countryside that
was made at a different time and
founded on different priorities. The
240 KV lines carried by steel structures marching in martianlike rows
across the land represent a scale and
a dependency on power never before
encountered. A more diversified and
smaller-scale system of power supply
would create less visual impact, and
much effort among environmentally

concerned organizations has been
dedicated to such alternatives. In
spite of the recognized need for appropriate technologies, the problem
today lies with the industrial forms
that symbolize current systems of
power and communication.

There are inherent contradictions
in attitudes toward the contemporary
landscape. We reject the new industrial forms while at the same time
finding beauty and appropriateness in
early and more primitive expressions
of technology, such as the windmills
of past European landscapes, the
railway, steam engine, and the mills
of early industrial textile production.
Our penchant for romanticizing the
past insures the survival of redundant
technologies. But it also generally

The rejection of new industrial forms is accompanied by the romance with old ones like the windmills that used to control water levels in the Netherlands.

A steam festival in Dorset, England. Old agricultural technologies, no longer a part of farming practice, are kept alive through preservation and festivals.

ignores the environmental degradation and human misery that accompanied the "dark satanic mills" of the nineteenth-century industrial revolution and which were abhorred by William Blake, Charles Dickens, Thomas Hood, and other moralists of the time. Yet these symbols of what Robert Thayer calls "Romantic Technology" (that which is no longer viable and has become part of history), "are critical to the visual/spatial vocabularies of public citizens as protection against the shock of the new and the ominous portents of the future. . . . we can process new information only by comparing it to what we already know."[2]

The Landscape as Urban Resource and Sink

The urbanization of the landscape takes many forms. The material and energy resources of quarries, mines, gravel pits, and forests that go to feed insatiable urban growth inevitably end up in used car dumps, sanitary landfill sites, and other places that receive the unwanted products of industrial society. Grady Clay, in describing what he calls "sinks," makes the interesting observation that topography is often a clue to social geography: "On the eastern shore of Maryland, atop gentle ridges and high ground sit the courthouses and big churches, the haunts of the whites; in

the creek valleys and by the sloughs one is more likely to find the blacks. The profile of vertical segregation is occasionally crystal clear. In most American cities, the richer you are, the less likely you are to confront the sinks; the poorer you are, the harder it is to climb out."[3] The need to hide everything we consider unpleasant, whether used car dumps, electrical transmission lines, or forest clear cuts, allows us to avoid the more profound implications of the regional imperative—the issue of environmental health.

The one-way systems in which the countryside both supplies the energy and materials as benefits to the city and subsequently receives its wastes as costly liabilities is the universal problem that links issues of identity and place with those of environmental health. Clay comments on the American landscape that "so noxious were many historic sinks that they became prime targets for public-health reformers in the nineteenth century, public housers in the 1930s, and urban renewers and highway engineers in the 1950s and into the present. In the process, hundreds of sinks have been filled and their residents dispersed, their old springs and streams bulldozed or piped underground with consequent loss of their scenic and biotic potential."[4] The process of "rehabilitation" of dumps, pits, and quarries leads, in the misplaced interests of environmental

The ever-present sink, often aesthetically safe behind fences, nonetheless avoids larger issues of environmental health and sustainability.

Metropolitan Toronto, light industrial suburb. As urbanization replaces agriculture, international horticultural landscaping replaces old productive land uses and indigenous plant communities.

clean-up and aesthetic standards, to the continuing dreariness of the landscape. Rehabilitation, now a legal requirement of mining in many parts of North America and elsewhere, ends up as a process of spreading topsoil over graded out slopes and introducing a green carpet and exotic lollipops over a site which, if left to nature, would have produced, over time, a new, biologically diverse environment. Rehabilitation of degraded habitat is often the featureless rural counterpart of much civic landscape. The problem is, however, not the *concept* of rehabilitation but the way it is carried out.

THE ROLE of design and aesthetics in establishing harmonious relationships between the old and the new in today's rapidly changing landscape is fraught with confusion and uncertainty about what its purpose should be. The industrial and technological forms that have become part of our countryside are a consequence of our way of life and are a necessary part of its future. We want the transmission lines for the power they bring, but we feel uncomfortable about the visual and ecological consequences of their presence. Considered in isolation, however, preoccupation with aesthetic values contributes to the increasing deterioration of the larger landscape. What is often masked are the underlying issues of natural diversity and connections to the land on which design must be based and

from which a new sense of the region can arise.

Farming, Forestry, and the Changing Landscape

As Nan Fairbrother has observed, it is industrial, not rural, societies that produce farm surpluses.[5] Yet everywhere in the developed world the earth's resources that sustain human life through its soils, water, and forests are in crisis. Old field patterns, developed in response to earlier determinants of farming, are disappearing in favor of fields scaled to suit modern machinery. Capital-intensive agriculture produces more crops on less land, but it takes more energy to produce the crops and the livestock than the energy returns to society.[6] And in spite of increasing production rates many countries have become net importers of food. Ontario, for instance, imports $3.6 billion of food per year. As Simon Miles has indicated, "today we have an industry in which . . . the competition of the market has pushed production upward by keeping prices low. This presents a very real dilemma for a society with an interest in the well-being of its children. . . . In order to enjoy our cheaper food to maintain a competitive position in international markets and to achieve other social goals, we have resorted to living off environmental capital."[7]

The basis for sustainable agriculture and regional diversity at a social level, the small "family farm" is becoming a thing of the past as international competition and the importing of cheap food from distant places forces more and more farmers off the land. The industrialized agricultural landscape—of field, hedgerow, and woodland—is not the diverse environment that existed in the days of mixed farming. The degradation of soil, loss of forest and wetlands, and pollution of surface and ground water have all contributed to the overall deterioration and reduction of landscape diversity. The new farm, as a glance at polder agriculture in the Netherlands shows, is an industrial environment, devoid of the varied habitats that the city dweller enjoys as a recreational experience. For the farmer such variety is simply an impediment to efficient farming.

Other patterns are emerging in the rural North American landscape. Much abandoned farmland, such as the marginal rocky landscapes of Ontario's Bruce Peninsula, is reverting to forest through natural succession, thus reviving its original identity. The abandoned cotton fields of Georgia have returned to native pine, and today forestry has replaced cotton as the major rural industry, radically changing the character of the Piedmont landscape. The Shawnee National Forest in Illinois consists of failed farms bought by federal agents during the Depression that have either been replanted or allowed to re-

vert to second-growth timber.[8] The New England farmland of a century ago has become the New England forest of today, with over 80 percent of the region's land area under tree cover.[9] This is in marked contrast with 1880 statistics when all states but Maine were dominated by agriculture, and it represents a return to the historical structure of land as it was used at the time of settlement by the Europeans.[10]

There was a time when land management meant the integration of farming and forestry, and where questions of natural beauty were the consequence of good land husbandry. These industries have long been separated at considerable environmental cost to the landscape. The dilemma of land management is rooted in the contemporary ethic of constant growth and the denial of a land aesthetic that would, if practiced, grow naturally out of environmentally sound management. This is central to the issue of regional identity, and is well illustrated in attitudes to forestry.

Forests and Aesthetics

Some years ago I spent considerable time touring the forest districts of Ontario and evaluating forestry practice. In one of the boreal forest management units my party stopped to look at a clearcut operation in progress, a vast area of cleared forest stretching as far as the eye could see.

Oregon. Landscaping the clearcut: aesthetics as the sinecure for beautifying forest destruction.

As we surveyed this dismal scene, a very large Paul-Bunyon-style forester from the logging company stepped up to me, and in a voice that was at once stentorian in volume and caustic in tone, uttered these memorable words: "So you're the guy what's been sent to plant flowers around our clearcuts, eh?" For a moment I was tempted to rejoin with well-measured and devastating sarcasm: "Yes, and with you six foot under." But another glance at Paul Bunyon's enormous bulk persuaded me to tone down my response to some mumbled banalities about aesthetic forestry.

Paul Bunyon's comment seems to represent, in varying degrees, an underlying contemporary attitude toward the land. It is very different from our understanding of traditional values where conservation is a necessary part of sustainable land use. There seems to be a widespread understanding of aesthetics as some kind of panacea applied to beautify the unfortunate but necessary scars created by the destructive exploitation of natural resources.

Aesthetic values, rather than enhancing regional identity, have somehow contributed to its disappearance. In the United States, increased clearcutting in the late 1960s and 1970s coincided with public concern over environmental protection. Opposition to such forestry practice came from those who believed, not unjustifiably, that it was associated with the ruthless exploitation of resources. The response among foresters was the creation of visual buffers so that people would be protected from the unsightly scars on the landscape. Hiding clearcuts became required practice, and forestry manuals were written to insure that the results of logging would look neat and visually unobtrusive. This was possible in flat landscapes but not in mountainous terrain where the efforts of "aesthetic forestry" to conceal cuts, or to give them a pleasing appearance, have been a major preoccupation of design manuals.

In many Canadian provinces, the establishment of what came to be known as the "Idiot Strip"—those uncut and unmanaged reserves along publicly traveled roads and waterways—has long been established policy. Their purpose was to protect the public from the visual impact of logging beyond the reserves. "Out of sight, out of mind," of course, is a motto that applies not only to forestry but to other land management issues. Along highways these strategic fringes of forest have become a visual no man's land, defying any sense of place or identity with one's whereabouts and an expression of current environmental values. The public, hoodwinked by a public relations device, is led to believe that all is well in the forest, that it is being safeguarded for future generations. This is in marked contrast to traditional European attitudes toward well-managed forests, such as the forested parks of

See-no-evil. The idiot strip hiding what lies behind.

Zurich, that are maintained in perpetuity and in full public view as productive forestry, as water and wildlife conservation, and as recreation.

Responsibility for forest recreation and aesthetics has traditionally been seen as the role of the designer. As the expert on visual impact, his stock in trade is to screen views, lay out campgrounds, and insure that forestry operations are well insulated from the recreation experience. Forest interpretation programs have generally ignored environmental impact from the exploitation of forestry and recreation, focusing instead on natural settings unhampered by human activities.

The severe and continuing depletion of forests throughout the world is well documented and has global environmental implications. In spite of all evidence to the contrary, our view of forests as a limitless resource still persists. This perception is maintained by the traveled route and the "idiot strip" that dictate the way we see the environment. There is a conflict between this environmental reality and perceptions of a world of plenty where conservation is a question of moral choice rather than necessity. In this cultural environment we have grown to accept aesthetics as a thing apart, a separate issue that can be applied as a cosmetic device to the landscape in general, and to the forest environment in particular. There is an unstated assumption that visual criteria can, or should, resolve land management problems, or at least provide them with a presentable face. All this emphasizes the dichotomy between appearances and the underlying determinants of landscape that give it form and identity.

HOW WE have come to these perceptions, particularly in North America, is complex. Even though there may be no definitive answers, it is worth exploring some historical clues that may shed light on why aesthetic values are so far removed from the sustainable management of our contemporary landscapes.

In many ways attitudes toward the forest landscape in North America have roots in a pioneering past and in perspectives imported from British tradition. At the end of the seven-

teenth century and into the eighteenth century, the natural forests in Britain were so depleted that the navy, Britain's basis for power and political strength in world affairs, was threatened by severe shortages of hardwoods, and this led to the reforestation of the country.[11] The success of forestry rested on the soundness of the agricultural economy on which the various planting schemes of the private landowners were based. It was, in effect, integral with agriculture as a self-sufficient rural economy, supported by government but carried out by the private sector.

The eighteenth century also saw the development of the English landscape garden—a rural movement that began with men of culture and taste. What was known as the picturesque style was derived from Italian artists who painted landscapes as a series of images imbued with great emotion and drama, and the early landscape gardeners felt that such art could also be expressed as landscape design. The influence of Shenstone, Kent, Capability Brown, and others was immense. They helped shape a new countryside aesthetic whose form was evolving as a consequence of economic factors—the need for agricultural and forestry reform—which were assisted by the Enclosure Acts. The changes to the countryside were not made, therefore, without regard for their good appearance.[12] Use and beauty became an integrated force that contributed to the country's agricultural and timber needs while providing harmonious scenery.

The nineteenth century brought the iron ship and industrialization and, with these, the decline of the old rural economy. Cheap food began to be imported from the American Midwest, frozen meat from Australia, and an endless supply of softwood timber from Canada. Industry, based in the urban areas, now provided the wealth required to maintain the country estates. Woodlands—now no longer a part of a self-sufficient rural economy—came to be prized as places for hunting and other pleasures.[13] As cities expanded, public interest in rural areas for recreation and relief from the unhealthy cities increased. Subsequently this became, in all industrialized countries, one of the prime uses of the countryside and has changed the nature of its economy and regional character. An emphasis on amenity, independent of the influences that create its visual appearance, has evolved with the changing needs of recreation.

This isolation of aesthetic perception from the function, ecology, and economy of the rural landscape was further aggravated by the emergence of scientific horticulture in the nineteenth century. Botanists, exploring the New World and the Far East, brought back thousands of new plants for cultivation and display in one of the century's innovations—the greenhouse. With the founding of the Royal Horticultural Society in 1804,

a scientific and aesthetic interest in plants as individual phenomena was born. They were to be admired for their peculiarities of form, flower, and capability for genetic manipulation. Horticulture replaced the homespun, rural knowledge of site, soils, and natural associations of plants. Plants could now be grown anywhere without regard for the environment where they were located. Carpet bedding, the lawn, the subtropical garden, and floral arrangements became the conventional expressions of landscaping and the beginnings of a style of design that had little interest in the inherent regional diversity of the vernacular landscape.

The English garden originated at a time in Britain when its forests and agriculture were being reestablished. Aesthetics and environmentally sound management had worked hand in hand within the framework of a rural economy to create a self-sustaining landscape of great beauty and utility. The horticultural tradition and the picturesque garden, together with the aesthetic that had been created by them, found firm root in North America. Imported to a continent where the clearing of forests, exploitation of virgin timber and soils, and survival were the primary motivations in dealing with the land and where a cornucopia of resource products was the basis for the economy, all that could survive was an artistic doctrine that had no roots in a land ethic. Making gardens was centered on already tamed, man-made environments with scientific horticulture providing the technology. It was totally dissociated as an aesthetic from land management as a biologically sustainable process based on practical necessity.

In addition, the very different forest types that occur in most of North America raise the issue of accepted images of sylvan beauty. Through generations of horticultural indoctrination we have an image of forests that is essentially urban in origin: an international cover of mature trees rising from a well-kept green carpet of turf with views unobstructed by messy undergrowth or regenerating plants. This image has permitted manuals of aesthetic forestry to suggest, for instance, that the "removal of dead branches from forest trees can contribute to a general appearance of neatness" that may have not existed before.[14] It is a kind of Disneyland stereotype that has no basis in natural process, succession, or differences in forest type. So one can sympathize with my friend Paul Bunyon for being cynical about any attempts to beautify his clearcuts. He was simply locked into a cultural view that has become a universal way of seeing the world.

IF CONVENTIONAL aesthetic values have little validity in land management, what can replace them? Another set of parameters is needed that will provide us with a relevant contemporary strategy for the regional landscape. There are several points to consider.

In the absence of an artistic or aesthetic tradition that is integrated with the objectives of land conservation such as occurred in mid eighteenth-century Britain, or in places where such values still exist, aesthetic criteria on their own do not contribute to better forestry or land management. In fact, the opposite is usually true. A policy of concealment by zoning or reserves tends to perpetuate bad forestry and is also bad aesthetics. In addition, it contributes to the deterioration of regional identity by maintaining an unchanging facade that is as dishonest as it is exploitive.

Aesthetic appreciation of the landscape is a subjective response to the forces shaping that landscape. As I suggested in chapter 2, it is a consequence of experiencing and of understanding landscape processes as opposed to the imposition of irrelevant design standards. Thus, sustainable, ecologically sensitive forest management does not need artistically trained foresters or, for that matter, aesthetically trained designers. Their role lies elsewhere, in the application of biologically sound management processes.

Integrating aesthetic values and biologically sound land management can be achieved by a process of landscape rehabilitation and renewal that recognizes a long-term investment in the land. Rural areas are developing a new role in their connection with the urban region and the industries and technologies that are reshaping the landscape will all have a stake in its continued health, even though many of these are currently not perceived to be "land connected."

Sylvia Crowe has suggested that the sense of chaos developing in the new landscape is partly because there is "no tradition of design to deal either with the new shapes, or with the landscapes which they create."[15] Although this is true, and some progress has been made in many places to blend industrial shapes with the natural features of the landscape, design by itself cannot deal with the basic problems of context. Underlying the visual disruptions to the familiar landscape are a number of factors that are relevant to these issues. For the future, design strategies are needed that will integrate the new urban region with its land base and help shape a productive and ecologically healthy regional landscape.

Some Opportunities for Changing Rural Landscapes

One critical aspect of the changing landscape is the issue of agricultural surpluses. Within the European community, the Common Agricultural Policy has been eminently successful in transforming Europe's food supply from one of shortage to surplus. A report by the Ministry of Agriculture,

Fisheries and Food in Britain has stated, "While serious cases of famine remain, the worldwide trend is clearly towards overproduction, not underproduction."[16] The volume of agricultural production has, in ten years, risen by over 20 percent despite a fall of 2 percent in farmland and 10 percent in the workforce.[17]

Overproduction has begun to shape policies that will have radical impacts on the regional landscape. Among these are opportunities for enhancing the health of a countryside that has suffered the devastations of industrial farming production methods. In addition, changing patterns of food consumption and a widening concern about the environmental consequences of deteriorating soils are other influences that will help determine what, how much, and in what way farmers will grow food. There is an increasing premium for food produced by "organic" farming such as free-range poultry, meat produced without steroids under extensive husbandry, and pesticide-free vegetables.

There is also the potential for new diversification of the rural economy as it becomes necessary to curtail rather than increase agricultural output. Policies in Britain, under active consideration in the late 1980s, were aimed at reducing production by 20 percent, with payments offered to farmers who comply for at least five years.[18] The implications for landscape change of such policies are significant. Among the options under discussion are the transfer of land from productive to amenity uses; the reforestation of marginal and arable lands to increase timber production and reintroduce amenity woodlands; and the reduction of pesticides and artificial fertilizers in a less intensive use of agricultural land.[19] Farmers within the Pennine Dales Environmentally Sensitive Area, for instance, may receive government financial support to continue farming in sympathy with the needs of the environment, wildlife, and the preservation of treasured landscapes.[20] France has targeted subsidies to specific valleys in order to sustain small farms and local handicrafts. The results are living museums that provide local income and preserve natural beauty.[21] In the 1960s the United States government responded to the grain surplus by offering modest grants to convert cropland to parks and conservation areas.[22]

The development of crafts and recreation in less agriculturally productive and naturally scenic areas in the Western world has become a significant rural land use. The social benefits brought by new rural wealth include new employment and the maintenance of traditional skills and crafts. Many a rural settlement in Ontario's lake country has gained a new lease of life from the influx of urban cottagers that provide local employment and income. Traditional crafts in Britain such as wall building, hedge laying, thatching, and re-

habilitating old buildings, have re-emerged to flourish in the enhanced economic climate of many rural areas. At the same time, the influx of urban people can also have an environmental cost if the process does not contribute to the health and sustainability of the land. The reduction of agricultural output and opportunities for new wealth to create environmental benefits provide a central focus for alternative land use policies and the establishment of a new sense of regional identity.

New Landscapes of Wealth

Between the expansion of North American cities and the urbanization of rural land, there is growing public concern about the preservation of irreplaceable productive soils. But methods to retain farmland have been successful only to the degree that they are part of a larger effort to enhance the economic viability of agriculture.[23] In a free-market economy one cannot separate farmland preservation from its economic support if it is to continue as a productive endeavor. Productivity, environmental health, and beauty have traditionally been the consequence of long-term investment in the land. The agricultural revolution of eighteenth-century Britain concentrated the ownership and occupation of land in fewer hands. The redistribution of land provided the wealthy squires with new sources of land, labor, and income, which they could use in profitable agricultural expansion. It was also accompanied by the work of the landscape designers, hired by landowners with a concern for the appearance of the evolving landscape. It was a landscape whose regional identity was fundamentally both new and determined by a combination of function, wealth, and beauty.

The famous Kentucky horse farms provide a contemporary parallel to this eighteenth-century landscape. The bluegrass region of central Kentucky is a unique combination of geology, soils, and climate. Underlaid primarily by thinly bedded phosphatic limestone, the soils are highly productive and traditionally have been farmed for row crops, hay, and cattle. Over the years, however, cattle farming has given way to horses since the mineral-rich soils, vegetation, and water are said to be the key to producing thoroughbreds with sound bone structure. These attributes, combined with a gently rolling pastoral terrain, a temperate climate, and the tradition of the English country squire has created a cultural landscape based on the horse that is the center for thoroughbred breeding in North America.[24] "Horse farming is a unique combination of big business, agrarian traditions and high society. Not the simple life of Arcady. . . . but rather the simplicity of ultimate sophistication."[25] The financial investment in the 40,000

horses that make up the bluegrass equine population are valued at close to $600 million.[26] Consequently, the investment in and design of the land itself to support and nurture such valuable animals is enormous and is specifically related to their functional requirements.

The shaping of fields is critical to preventing accidents. Sinks, potholes, and steep slopes are dangerous to running horses and contribute to injury. Flat fields can be wet and destroy turf. Fences, usually made of painted or treated oak, are doubled to separate animals physically into different fields and to circumvent potential hazards such as trees or rock outcrops; rectangular corners are rounded. Barns, which represent major financial investments, must be located, designed, and ventilated to minimize the danger of infections. These and similar requirements set the conditions for the organization, management, and form of the landscape. In addition, however, they provide the basis for a model form of agriculture. Pastures are kept in a stable and productive condition by mowing and chain harrowing to reduce unwanted "weed" vegetation, since few owners risk the health of their animals from the application of pesticides and herbicides. Soil amendments to the pasture are applied as needed in small areas, thus avoiding the stream and lake pollution that is inherent in the excessive fertilization of farm fields.[27]

The Kentucky Horse Farms. A unique contribution to landscape health combined with wealth creates a beautiful landscape. Here, the estate residence is set in spacious grounds.

Land management and design are integrated. Lakes and fences are carefully sited to conform to the lie of the land, and to combine functional and aesthetic purposes.

The high capital investment of a horse pasture and the need to maintain its productivity bring a level of beneficial management that sustains the health of the land. The healthiest horse is the one whose environment approximates nature most closely. In fact, the emphasis is on creating environments best suited to horses rather than to people.[28] The development of the bluegrass horse farming landscape has once again brought the skills of the designer and land planner to bear on the shaping of these modern, wealthy, rural estates. Like the eighteenth-century landscape designers before them, they have begun to integrate the functional requirements of animal husbandry with an understanding of landscape tradition and visual relationships. There is a harmony in the bluegrass landscape, between topography and fence alignments, between tree planting and new lakes; in the enhancement of views and respect for the shape of the cultural landscape that preceded the modern farms; in the location of barns to reflect traditional hilltop sites. Traveling through the countryside one is struck by the inevitability, serenity, and sense of care in this contemporary scenery. It has been created by the economic realities and environmental requirements of a prosperous rural industry and enhanced by design—a landscape of prosperity, environmental health, and distinctive regional character.

Technological Form and Rural Patterns

Industrial forms that have been superimposed on the landscape appear foreign to the traditional cultural patterns and environmental character to which we have become accustomed. Several reasons underlie this purely emotional reaction to change.

Most industrial objects are technological extensions of the city, not the outgrowth of rural imperatives. By contrast, where technologies have developed specifically in response to rural needs, the resulting forms have a greater likelihood of belonging, of being part of the landscape. For instance, few people object to the prefabricated windmills that used to be associated with every farmhouse for drawing water, or the waterwheels that at one time provided power for the local sawmill, or the contemporary industrial silos that store the grain. They are accepted as part of the rural scene because their functional relationship to the necessities of farming gives them a clear sense of belonging to the agricultural landscape.

Technology has created objects whose construction and materials are derived from the discipline of that technology and its manufacturing processes. They have, therefore, no links to the native materials of the landscape itself. Thus, there is little obvious visual relationship to the physical landscape in the way that was apparent in early agricultural technologies. The early preindustrial stone barns, walls, and hedges of Europe, or the wooden structures and zigzag cedar fences of Ontario, were the products of their landscapes in function and materials, which reinforced their regional character. Their forms derived from the limitations of the materials available and the most practical way of achieving what had to be done. The same imperative is true today. Except in cases of historic preservation where special grants are available, no farmer interested in staying economically solvent will build a stone wall or lay a hedge, when barbed wire and wooden fence posts are cheaper and will keep the sheep in more effectively.

When industrial objects have nothing to do with farming processes the denial of regional character is all the more apparent. Installations such as industrial plants, electrical power grids, or gas and oil pipelines carrying power or energy generally contribute nothing to the environments in which they are located or through which they pass, either in terms of promoting productivity or of environmental health. At best, their location avoids major environmental damage; at worst the opposite holds true: industrial processes leave a legacy of pollution and degraded landscapes, and the management of rights-of-way is often destructive of vegetation and wildlife habitat. There is no investment in the place.

To establish a basis for making meaningful connections requires that industrial processes must *contribute* to the landscapes they change through a commitment to their ecological diversity and an enriched cultural environment. In doing so they can provide a relevant basis for appropriate design connections to their landscapes. Thus, although efforts have been made to design industrial forms that are more pleasing and less obtrusive on the environment—and some have superb engineering sculptural elegance—visual concerns are the stuff of changing values and taste. How things look is much less significant than their environmental relationship to the land. A basic principle of a new regional identity is, as I have suggested, one of sustainability and environmental enhancement. The corridors required for power transmission, for instance, are often lands sterilized by herbicides, or made unproductive as turf deserts, and do not benefit environments through which they pass. Yet the potential of such land corridors to provide benefits is great. Permitting rights-of-way to naturalize as distinctive wildlife corridors, or to be used as productive market and allotment gardens, are precedents that can occur in many places.

Industrial Production and the Enhancement of Nature

A significant example of environmental benefit in Britain is the Central Electricity Generating Board's field study centers and nature trails that have been created on substation sites.[29] Since the early 1960s a policy of using nonoperational land for public benefit had been advocated. It was proposed that public use would be considered in the context of local needs, suitability of the land, cost, and similar factors. By the mid 1960s conservation became government policy and the Midlands region of the Central Electricity Generating Board was the first to acknowledge that it might have a social responsibility beyond generating electricity. In 1967 the first nature trail, field study center, and nature reserve was opened. Since that time a number of others have followed, developed in association with county councils, education authorities, and naturalist societies. Many of these have included existing woods and ponds, while others have been purposefully created from derelict land. Each has been developed to reflect and enhance local site conditions and regional characteristics. For instance, one involved the rehabilitation of old water-filled gravel workings, another was associated with a disused railway line which was allowed to regenerate, a third was developed on a site containing a stream, ponds, woodland, and water meadow. In each substation site the Central Electricity Generating Board has provided a field study center and classroom building, and nature trails that are used by local schools for environmental education.[30]

The enrichment of the environment can also be realized and sustained by other major land-based industries. An example of this is Petro-Canada's oil refinery in Clarkson, Ontario, whose environmental conservation measures in water quality, natural habitat creation, and historic preservation have been in effect since the mid 1970s.[31] Surrounded by residential and commercial development on three sides and by Lake Ontario on the fourth, the refinery initiated a landscape rehabilitation plan in 1975 as part of their industrial expansion program. Undertaken in cooperation with its residential neighbors, the program included an extensive buffer zone of planting and wildlife habitat creation that formed a natural link between an existing marshland (the Rattray Marsh conservation area) on the lakeshore and a wooded creek extending up to the refinery's easterly boundary. The reforestation of its boundary effectively completed a natural circular link connecting the wetland, creek, and lakeshore, and this has permitted free movement of wildlife and the reestablishment of indigenous plants that had all but disappeared.

A nature trail attached to the Central Electricity Board's 400KV Pelham substation, situated on the borders of Essex and Hertfordshire, England. Where industry contributes ecological, productive, or community benefits to the landscape it uses, environmental health and identity are enhanced. Industrial objects may thus become a part of, rather than an imposition on, the landscape. (Central Electricity Generating Board, London)

Part of the Sutton Courtenay nature reserve at Didcot, England, includes a field studies center and a fifteen-hectare nature trail. Preservation of original trees and a massive revegetation program were part of the landscape program. Some 2,000 children visit the nature reserve in Berkshire each year. (Central Electricity Generating Board, London)

A plan for rehabilitating the periphery of Petro-Canada's Clarkson Refinery in Mississauga, Ontario, included the creation of forested connections with a remnant marsh on Lake Ontario and establishing natural areas behind security fences. In this way a heavy industry contributed to wildlife diversity and a rehabilitated neighborhood environment. (From Hough, Stansbury + Woodland).

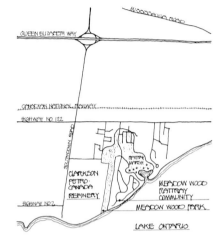

Part of the forested corridor along the refinery's edge, managed as a natural area.

Since its inception in 1975, the refinery's landscape has been managed as a wild area. Initial planting of various woodland types included groves of white cedar, spruce, and deciduous associations that now form distinct habitats. In combination with the extension of a small wetland, a pond, and a meadow, a complex series of natural environments has evolved in an area of less than twenty hectares. The management plan called for some thinning of tree groves over a period of years to encourage healthy growth and the establishment of a canopy and ground flora. Some periodic mowing along residential boundaries established a connection with the well-kept public landscape while maintaining the remainder in a natural state to encourage regeneration. Over ten years a diversity of new wildlife has recolonized the area. Muskrat have appeared along with a den of foxes, breeding Canada geese, ducks, and countless songbirds. And in 1985, deer arrived and took up permanent residence within the re-

One of the deer that have taken up permanent residence within the refinery. The natural areas and their wildlife now form part of public tours by refinery personnel. (Reno Masaro)

finery boundaries, leaping over the two-and-a-half-meter security fences and taking advantage of the wooded habitat and cedar trees as a winter food source. For the residents surrounding the property, the presence of such abundant wildlife on their doorstep has created a sense of wonder and fascination. Neighbors put out saltlick for the deer; refinery workers leave their left-over sandwiches; groups of children being shown around the oil production processes are taken to the nature reserve to get a glimpse of wild deer living in a refinery.

Other facets of the refinery's environmental management include a carefully controlled water-treatment system that insures that lake water, used as a coolant in the refining process, is returned to the lake in the same condition as it left. Originally located on the lakeshore, a historic house was permanently moved to property donated by the company and provides an opportunity to interpret the historical and cultural heritage of the urban area. Regular meetings with the community, established during the initial planning phase, maintain dialogue to insure that issues of environmental quality are resolved before they become problems.

The significance of these examples is that major industrial concerns have demonstrated their belief, first, that industry has an investment in a healthy environment and in natural and cultural history; second, that industrial complexes can contribute to and enhance the regional landscapes they occupy; and third, that they can contribute to an understanding and knowledge of ordinary local land-scapes. Finally, the benefits that accrue to industry itself are a heightened public awareness of its function in society and its role in environmental conservation.

Integrating Urban and Rural Landscapes

The technological and social changes taking place today suggest that the conventional dichotomies between city and countryside could give way to an integration of the two and begin to shape new landscapes and planning policies. Dorney has proposed a four-zone system in the urban region that includes large-scale farming with a single crop focus in the rural region; smaller family farms growing mixed crops in the urban shadow commuting zone (many Mennonite farms in the Waterloo Ontario region are an example of successful farming at this scale); small cash crop acreage at the urban fringe; and highly diverse, small-scale "backyard" gardens in the city.[32] At a local scale, however, other examples are beginning to show the creative alternatives that are available.

The changing social landscape has numerous facets. There is evidence that farm employment is growing in some rural areas in the urban shadow, an indication that near urban farms are economically viable. There are many newcomers to farming near

cities in North America and Britain whose principle income is derived from sources other than farming but who, nonetheless, contribute significantly to farm output. The preservation of productive farmland within and on the edges of cities in Europe has long been a fundamental part of urban planning policy. The vineyards that occupy the hillside slopes of Stuttgart are an example. The changing nature of the suburbs—from the domain of the Anglo-Saxon, white-collar worker to an ethnic population of high cultural diversity—is also adding to the number of part-time farmers. Many are moving from the traditional downtown areas, but others come directly to the suburbs from their native countries. Many bring with them a rich urban tradition of productive gardens and allotments that is transforming hitherto conventional and sterile suburban landscapes. Allotments spring up in waste places, under electrical utility rights-of-way and unused railway and industrial lands. Backyards become productive minifarms.

The shift in the suburbs toward higher densities and the opportunity to grow local produce by their less wealthy, new residents create opportunities for a more efficient and productive use of space. The market for pesticide-free produce close to the cities is growing among the general public and restaurants, which are often prepared to pay premium prices for quality fruit and vegetables over industrially grown food. Community gardens in American, European, and British cities have proved to be a lasting and powerful force in neighborhood change. Community-controlled open spaces are becoming a standard feature in cities in the United States,[33] a movement that is contributing to their future form and identity. In addition, the integration of the landscape of power and industry provides opportunities for generating a new contemporary landscape form.

Urban Industry and Rural Continuity

We have already seen how new development can contribute to the productivity of the land it occupies. It lies at the heart of a new identity for the evolving city's edge. High-technology industry frequently locates on the edges of cities because land taxes are low and large amounts of land are available. This kind of development has traditionally done much to degrade the essential characteristics of the existing landscape by ignoring its previous uses, and the patterns of its boundaries, hedges, woodland, and streams. Where development has contributed to the continuity of the regional landscape, identity can be maintained. The following example illustrates this idea.

When the office manufacturing plant of Herman Miller Incorporated of Atlanta, Georgia, acquired its new

site—a well-managed productive farm of 135 acres (54.6 hectares)—a decision was made to continue the farm operation that had been in existence since the early 1900s.[34] A phased management plan was initiated that included a return of the site's 41 acres (16.6 hectares) of pasture to productive use. At different times the land had been planted with a variety of crops including cotton, corn, tobacco, and soybeans. The new system for pasture grazing included cattle forage requirements, annual pasture, and hay needs for livestock. The site's forestland, which included a pine forest and a mixed pine-hardwood forest, was managed to produce revenue and recreation opportunities. Two existing ponds that supported various species of fish provided water for livestock and acted as detention ponds for storm water and run-off. They also attracted wildlife and migrating birds and insured visual enhancement. At work were the design principles inherent in the regional imperative, including economy of means, maintaining the history of the land while adapting it to modern purposes, sustaining the land through intelligent management, and the concept of investment in the land. They showed how contemporary change, far from creating placelessness, can provide an opportunity for diversity and a continuing, though modified, landscape identity. But in addition, the difference between this example and

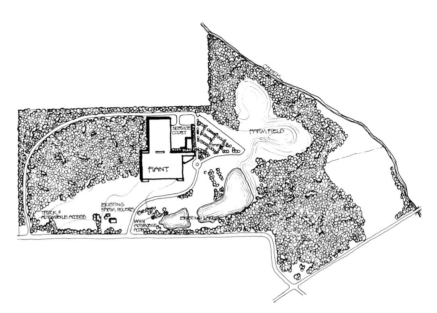

A view of a pond and meadow. The two ponds provide flood control and on-site water sources for livestock, wildlife, and gamefish. (Heery International, Atlanta, Georgia)

The establishment of Herman Miller's manufacturing plant on agricultural land brought technology to Georgia's rural scene. Surrounded by a working farm with pastures and woodlots, which it preserved and managed, the factory became a part of its landscape. (Plan from information supplied by Heery International, Atlanta, Georgia)

the typical industrial developments found so often on the edges of cities is one of process. The latter is seen as a finished, fixed product, a design imposed on its site, denying its landscape context. The former represents a process of evolution and continuity, one that derives benefit from and contributes to the land it occupies.

The new sense of place that has been explored in this chapter is also inextricably linked to a modern industry that thrives on the picturesque landscape and which has become a central force in the economy and survival of every community—that of tourism.

7 / Tourism

Searching for the Differences

Tourism has the potential to be a major force in the protection and maintenance of regional character. But like any other economic development, when the environmental and social values on which it depends are absent, the rich diversity of the natural and cultural landscape is degraded and somewhere becomes anywhere. In this chapter I examine some of the issues of identity in relation to this fast-growing and significant phenomenon affecting the contemporary landscape.

A Place in the Sun

Anyone who has spent a holiday in Blackpool, Bournemouth, or any similar British coastal town will know just how miserable they can be. Among my summer memories of these seaside resorts are people huddled in deck chairs on the beach, staring stoically out to a lead-gray sea, wearing heavy overcoats against the cold, and holding umbrellas at the ready against the unremitting rain. Of course, this was before the days when people could travel to the Mediterranean and places with more reliable sunshine for their holidays; when being jolted from sleep at 6:00 A.M. by the strident call to "wakey, wakey rise and shine" was how one spent two fun-filled weeks away from home at Billy Butlin's holiday camp. That was the choice there was.

My first and most vivid childhood memories, however, are scenes from the south coast of France on the Côte d'Azur. My two uncles made their living on the land as carnation growers, and in the 1930s the area was known for its thriving carnation and flower industry. The south slopes that overlooked the Mediterranean were well suited to this kind of agriculture. It was a memorable and intensely beautiful landscape. From the vine-covered arbor of the living terrace on my uncle's estate lay the pattern of fields—a geometric maze of stakes and string lines and overhead irrigation pipes providing support for the tender plants. The scene was a dramatic and overpowering combination of color from the red-brown earth, the white stucco walls and red roman tiled roofs of the farmhouses that terraced down the hillsides to the sea, the brilliant greens of the vegetation, the reds of the bougainvillea, and the intense blue of the distant sea and sky.

For my uncles, who had invested money and many years of labor in building up a successful business, the climate, soil, southern orientation, and slope were the essential ingredients to profitable and sustainable farming. For the visitor, the sun, dry air, sea, intense color, and unique scenic character of the cultural landscape made it an irresistible holiday resort. Indeed, that is what it was along sections of the coast: a place for elderly couples escaping the rigors of gray and inhospitable winter

The Côte d'Azur in 1980 from the terrace of my uncle's carnation farm, above what was once the small village of St. Laurent du Var, near Nice. Endless tourist development has radically changed the view of what was once farmland.

climates; a haven for artists and writers seeking inspiration in an intensely dramatic landscape of sea, sun, and brilliant color; a place for the rich and beautiful, where bikinis, skin, voluptuous and hedonistic indolence were what came naturally. In the 1920s and up to the Second World War, these two quite opposite styles of living coexisted. Flower growers farmed and made a living by serving the ever-fickle Parisian market and the tourists came and went. The inevitable outcome of the conflict between the farmer who used the land as a place of work and the vacationer who saw it as a place of leisure did not occur until after the liberation of France in 1945.

I visited my family home again in 1979 and found the region transformed. Where I had once looked over a terraced rural landscape of fields and small villages to the sea, there now lay an endless vista of hill-side vacation homes and apartment buildings, the old flower-growing estates mostly sold to the developers. Indeed, one of my uncles succumbed to real-estate pressures and sold out. On the coast there rose the latest in vacation condominiums, vast, terraced, pyramid blocks providing all the built-in recreational needs for the summer vacationer: marinas, tennis courts, swimming pools, boutiques, beauty parlors, and banks. The villages now formed a continuous urban conglomerate along the coast, the roads a nonstop traffic jam of honking overheated cars and sweating drivers from Cannes to Monaco; the hotels sporting the required palm trees as reminders to the tourists that they were indeed in an exotic place and not just anywhere. The old rural, working landscape has been edged out both by tourist development that has all but eliminated the very regional character that people came to

Condominiums for the French tourist: everything on hand for the summer vacation.

The Promenade des Anglais, Nice. Along the Riviera coastline there is now unbroken development and nonstop traffic. Palm trees and sunshine are among the few natural elements that remind people of their location.

enjoy and by government taxation that discourages sustainable cultivation. The sun and the sea remain to remind one of what was once there.

Benefit or Liability?

Once the exclusive purview of the rich, the summer vacation has now become available to anyone seeking sun, sand, sex, or simply somewhere new and different. In 1979, for instance, tourism in the Caribbean numbered some 7.5 million people.[1] Spending by tourists in several Caribbean island states exceeded 20 percent of the gross national product in 1977; in the Bahamas it was nearly 55 percent, and Barbados earned nearly 90 percent of its foreign exchange from tourism.[2]

Paul Wilkinson points out that many small countries that lack power and wealth, such as island micro-states, may have little control about whether to encourage or discourage tourism. The decisions are often made by other countries, multinational companies, or airlines. Because of their resources and commercial power, these international organizations create a situation of "comprehensive dependency." For example, "There have been few, if any, cases of an island government blocking tourism when an international airline decides that it wants to include a particular island on its regular schedule. This . . . has been clearly demonstrated for a wide variety of islands, including the Cook Islands and other South Pacific islands."[3] For cultural reasons, though, some Third World countries have adopted policies to discourage tourism. For instance, the remote Himalayan kingdom of Bhutan has placed a virtual ban on outsiders in the hope of preserving its Buddhist culture. That country decided that the annual profit that tourism brought in the past was not worth the accompanying problems, which included theft of sacred relics, desecration and looting of monasteries, and corruption of the local population.[4] For the most part, the perceived benefits, and attractions of tourism for countries that are beautiful, culturally rich, but resource poor, lead, in the absence of local control, to solutions that often have severe social and environmental repercussions.

THE FRENCH Riviera symbolizes an ideal in tourism. Yet its problems are typical of its larger region. Modern growth has turned the entire Mediterranean Sea into a dumping ground for industrial and municipal waste. This condition is exacerbated by the fact that the area attracts about one-third of all international tourists, most of them during the months of June, July, and August. In 1973, for example, almost 60 million people from around the world visited the region.[5] The Riviera's popularity, therefore, has contributed to crowding and to the pollution of its air, sea, and beaches.

While tourism clearly has economic benefits, the destruction of a region's scenery and native landscapes from concentrated use is often assumed to result from a lack of planning. Yet efforts in the 1960s by the French government to plan tourist resorts in undeveloped areas in the south of France totally transformed the natural environment into a man-made one. The resorts in Languedoc-Rousillon on the sparsely populated, low-lying coast west of the Rhone Delta wiped out the wetlands and flood absorption area that supported marine life and birds. They were replaced by a planned string of resort cities separated by "green" areas whose "highrise, city-like atmosphere . . . attracts a high style international yachting crowd."[6]

Planning, in effect, is no sinecure for a harmonious relationship between development and the environment. The problem rests with "top down" planning that ignores local values and knowledge as well as significant environmental conditions. The opposite often holds true when the regional identity of a natural landscape and the ways of traditional culture stand in conflict with the opportunities that tourist dollars can bring to a poor people. For instance, the impact of enormously increased travel to Nepal over the past twenty years has altered the economy and ecology of the area and disrupted the traditional way of life of the Sherpas.[7] Deforestation of the mountain slopes and the consequent erosion of the land are among the region's most serious environmental problems. These problems have been generated partially by the local pressures of a growing population, but tourism has dramatically increased the demand for firewood. Visitors want light, heat, cooked food, and hot showers to maintain the basic comforts they are used to at home.[8]

For the traveler, the essence of tourism is being away from home; for the tourist operator, the traveler is a visitor who spends money. It can be argued, therefore, that tourism engenders a lack of commitment to the place. As a form of recreation it feeds off the natural and cultural environment without returning any environmental or cultural enrichment beyond what is spent in casinos, gift shops, hotels, and restaurants. The international hotels lining the beaches, the swimming pools that replace the sea, the cuisine that insures as little deviation as possible from what is familiar, and the decorative resort landscaping that ignores local plant communities create a similarity of experience abroad that is insulated from the nature of the regional landscape.

Urban Landscapes and Cultural Fragility

The sensitivity to change caused by tourism varies from place to place and from culture to culture. As Fred Bosselman points out, this is particularly true of cities. A city with a diverse and dynamic image, one that is growing and evolving, can absorb mistakes that seem tragic in a city with an image of timelessness and solidarity.[9] He compares the development of hotel accommodation built to promote tourism in London and Jerusalem. The worst examples of hotel building have detracted from the peculiar regional qualities that make Jerusalem special. Its urban environment of native stone, massive building, and its skyline of mosques, churches, and temples is extremely fragile.[10] By contrast, the overall impact of hotel building on London, already a mish-mash of building styles and materials, is far less damaging.

The irony of culturally rich but economically destitute urban places is that their very economic lifeblood depends on the wealth that tourism brings, while at the same time it degrades the cultures that made them special. The migration of visitors from northern, developed countries to southern, underdeveloped ones has been increasing since the 1950s, and now some 50 million people travel to Third World destinations each year.[11] "North-South tourism injects the behavior of a wasteful society in the midst of a society of want . . . Tourism may act as a form of advertising for a modern consumer society, but to an audience that is generally too poor to afford it."[12] Although this has probably always been a characteristic of tourism, its impact is now felt universally since tourists are numbered in the millions rather than the thousands.

Advertising promotes the idea that one can travel to secluded, unspoiled places off the beaten track, historic cultures unsullied by Western influences, and still enjoy all the comforts of home. "Travellers today are discovering the ancient realm of the Queen of Sheba in what is now the Yemen Arab Republic. They can make a round of sights not dissimilar to those of the fifteenth century and still end their day in a Jacuzzi whirlpool or an air-conditioned bar. The international hotels have come to Yemen and with them travel agents and tours."[13] To an observer interested in indicators of change there are numerous tell-tale signs of how tourism is affecting such places. Among these are recreation yachts that outnumber working boats in the harbor; horse-drawn carriages full of tourists; old people in horse-drawn traps and the young on motor scooters; the absence of new development other than tourist hotels and condominiums; tour buses drawn up outside old churches; coca cola signs.

Design has as much to do with the homogenization of places as the tourist industry. The local vernacular buildings that respond to the practical necessities of solving problems are all too frequently replaced by designs that have little to say about the places where they are located. The designer's stock in trade is the universal landscape derived from years at schools that have traditionally promoted internationalism rather than regionalism. It is derived from working in every place but understanding no place. Eckbo observes, however,

that it is difficult to place all the blame on designers when the telephone and the jet plane make regional forms accessible to the whims of clients everywhere. "The tourist arrives in our [local place] from Anywhere, seeking he knows not what . . . How do we make Anywhere a Place?"[14] As John Betjeman warns, referring to the British scene:

> In a few years this country will be looking
> As uniform and tasty as its cooking.
> Hamlets which fail to pass the planners test
> Will be demolished. We'll rebuild the rest
> To look like Welwyn mixed with Middle West . . .
> We'll keep one ancient village just to show
> What England once was when times were slow.[15]

One of the dilemmas of travel to underdeveloped countries is the fact that the Western tourist seeks the differences of culture while condemning the fact that his presence has created a cultural impact. In fact, as Barrie Greenbie observes, "Almost anyone in any country deplores tourists when not actually engaged in being one himself, or profiting from their presence."[16] How often have we heard the comment from fellow tourists that "this place will soon be spoiled, so it's lucky we came this year." The increasing wealth tourists bring changes local styles of living that are often inconsistent with their image of colorful scenes of local markets, people, and street scenes—an image that often reflects poverty. Regional identity is something that is experienced by the tourist as something to enjoy. But for those who create that ambience in the first place, the feelings are very different. They dream of the places that we take for granted: the green spaces and gardens, the shopping malls, the spaciousness of the North American suburb. Necessity and opportunity dictate the nature of a place and its culture. When opportunity presents itself for doing things more easily and with less effort, change will occur. Modifying the destructive aspects of tourism, therefore, may be a matter best resolved by understanding the culture and environment of one's own home place.

Images of the Past and Fossilized History

Tourism thrives on the past. The enormous number of visitors who are attracted to forts, mission stations and colonial villages, citadels and old gun emplacements is an indication of why tourism is one of the fastest growing service industries in the world. Every town that has seen better days and can boast of a past has an eye to capitalizing on its own historic specialty for tourist entertainment and dollars. The process is lucrative though costly, for the essence of what can be sold is cultural and natural identity. Relph observes that regional differences are stylized into the cute and kitschy tourist attraction—the colonial South is embalmed at Williamsburg and colonial New England in Sturbridge Village.[17] And although the events of early settlement days were real and the places our forefathers made were direct responses to the environment and conditions with which they had to cope, there is nothing in the fossilized relics that have been painfully reconstructed, often at public expense, to indicate either continuity with the past or connection to the place. The entire sterilized setting of neat reconstructions, mown lawns, horse-and-cart rides, and all the other aesthetic pretensions created for the entertainment rather than the enlightenment of the visitor are guaranteed to discourage any expression of the local landscape of the time. An enthusiastic travel article on Fort William in northern Ontario describes it as closer to an early prototype of a shopping mall than a nineteenth-century fur trading post.[18] This is an apt, though unwitting, comment on the fact that most historic restorations are as physically isolated from their surrounding natural and working landscape as the shopping mall of today is from the communities that support it.

One finds little reference to ques-

Ste. Marie among the Hurons, Ontario. Fossilized history: reconstructed forts neatly packaged with landscaping and mown turf lack cultural or environmental links to the past.

tions of health and disease, unpaved streets, human waste, the enormous effort in clearing forests for agriculture, the subsistence living. These have to do with the people's way of life and environment as they really were. James Fitch discusses the question with respect to the historic town of Williamsburg, Virginia. He says that "we are confronted with serious discrepancies between the present image and historic reality . . . Williamsburg, as the capital of an agricultural province would have been a sleepy, seedy little town with a large population of black slaves . . . Its life-support system would have been primitive and unaesthetic . . . Street life would have been lively: cattle market, slave market, jail house, chain gangs, would have filled the streets with their sights, sounds and smells . . . the sheer sensory stimulation of living in Williamsburg would have been a powerful, if not always pretty experience." [19] In today's reconstructions, places like this represent serious sensory deprivation and

misrepresentation that some countries have attempted to rectify, sometimes going to extraordinary lengths to do so. The Dutch, for example, have a historic farm educational program for the benefit of school children, in which the government pays the farmer to continue old agricultural practices that have been abandoned.

Whether history is represented by forts, painstakingly put together through unimpeachable records, interpretive centers explaining great events through artifacts in glass cases, films and old photographs, or pioneer villages, carefully assembled and neatly packaged with landscaping, the results for regional identity are fundamentally the same. The cultural and environmental links are gone. They remain empty shells cast up on the shores of the present—objects isolated from the processes that shaped them. The attraction that such relics have for people is undeniable. The visitor pays his money, stares, and leaves. The adage that today's garbage is tomorrow's archaeological dig appropriately expresses the essential dilemma of the regional imperative. For the builders, farmers, and artisans, living their daily lives and making the best from what they had, keeping and adapting what was useful and throwing out that which was not, made simple common sense. Today their buildings, tools, and artifacts have an entirely different kind of value for the collector or the tourist who admires them in isolation from the conditions that gave rise to them. It is aesthetic and scientific, rather than utilitarian priorities that prevail.

Continuity and Heritage

As I suggested in chapter 3, what distinguishes vernacular towns and villages is a sense of continuity. Throughout history buildings, squares, old walls, and paving stones—the fabric of urban form— have continued to be adapted to the conditions of the present. It is part of the process of living. Every old English parish church bears the mark of continuous modification and addition dictated over the centuries by need, political events, and social change. Much of what is experienced by the visitor who wanders through towns and cities, therefore, is adaptation and change through time and in the context of the overall social and physical fabric.

Much of the philosophy of architectural preservation has been based on the conception of old craftsmanship and buildings as museum pieces. The wholesale destruction in the 1960s of the urban context surrounding the historic buildings of the Independence Hall area of Philadelphia, for example, is typical of the values of the time which, in some ways, still prevail. Pierce Lewis has observed that nostalgia is a growth industry, yet despite the escapism that much historic preservation implies, "huge numbers of intelligent and affluent Americans long to be rooted in a world of permanence, . . . a world wholly at odds with the progressive idol which [they] are all supposed to worship."[20] There is, however, increasing evidence that historic preservation also means adaptive reuse and continuity, not only of buildings, but of entire sections of the urban fabric.

The process and continuity in the social and cultural fabric of urban places have parallels with the cultural landscape. Expressions of regional identity, although basic to built form, are also fundamentally part of the cultural landscape that preservation values often ignore. All man-made landscapes, including the great private estates, botanical gardens, parks, and particularly working farms are tied to their geographic, climatic, and historic context. They represent time and place and must, to stay viable, have relevance to life today.

As James Fitch correctly observes, very few historic landscapes have a physiognomy that corresponds to the way their designers originally conceived them.[21] Some do exist, however. The great gardens of Japan were made and nurtured over centuries as perfect works of art, and have changed little. Some gardens in the French formal tradition—Versailles and Vaux le Viconte, for instance—have been maintained in the way that they were conceived. But for the most part, in the absence of constant attention and of the controls necessary to the classical tradition, these gardens have long lost some of their formal appearance and have been softened by years of neglect. Their history, like the history of most landscapes, is linked to the evolution of plant communities whose appearance and shape alters dramatically over time. For example, a comparison between Etienne du Perac's famous 1573 engraving of the Villa d'Este near Rome and the present gardens clearly shows how nature has taken over and transformed them into a magical place of intense regional character—not what the designer intended but what nature has created. Much discussion in the literature of historic preservation focuses

**Tivoli,
Villa d'Este**

*The garden as
Ligorio intended,
from an engraving
of 1573 after du
Perac. (Dumbarton
Oaks, Trustees
for Harvard
University)*

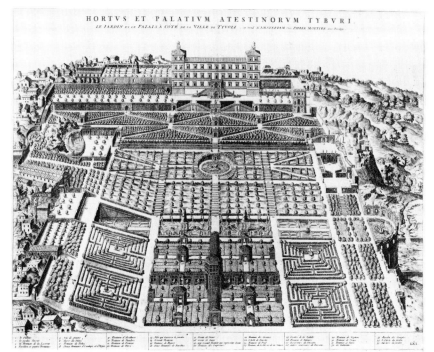

The Villa d'Este today as nature intended it. Here, the main axis focusing on the villa.

The Pathway of
the Hundred
Fountains, today
overgrown with
maidenhair fern
and moss.

The Fountain of
Tivoli. A riot of
vegetation enriches
its character.

A stone carving from the Hundred Fountains.

on how to sustain such places, the factors involved in restoration with respect to their original form, botanical and horticultural accuracy, and the period to which they should be restored.

Of course, there is obvious validity in treating some places as historic museums, particularly the great palaces, theaters, temples, and gardens of past cultures. These expressions of great art that Rudofsky describes as the "Pedigreed architecture of power and wealth,"[22] naturally fit the historic preservation mold to be seen, admired, and studied. The problem comes when change and adaptation to new conditions are precluded from the process of preservation.

Ethnic Culture and Conformity

Once a North American phenomenon, but now universal in the Western world, ethnic cultures are a fact of life in cities. Their presence provides a large part of the vitality, diversity, and identity in the contemporary urban scene. Chinatown and ethnic markets have also become major tourist attractions as colorful, interesting, and different places where one can still find poultry hanging from hooks in the butcher's shop, cheese shops, bakeries, fishmongers that sell octopus and squid, places that cater to the special needs of a diverse ethnic population, and a whole sensory experience of sounds, language, and smells. The phenomenon of being in Hong Kong, Italy, or Portugal has now been transported to one's home territory, where, for the cost of a streetcar ticket, local residents and visitors can see the world. Everywhere brightly painted houses, flourishing gardens of sunflowers, bric-a-brac, religious icons, and fences express ideas of personal territory and a rich urban tradition. Over the backyard fence one will find productive, urban mini-farms, with grapevine-covered trellises providing shade and a future wine harvest.

Ethnic neighborhoods reflect the survival of rural and urban skills and adaptation to a new way of life. They have evolved as a cultural vernacular, adapting the gridiron of the city's fabric to fit their own needs and cultural expression. Chinese, Portuguese, and Greek communities have special identities that are microcosms of their countries of origin. As their potential to enrich the city and attract tourists is recognized, their preservation as communities begins to have some significance, but this potential also brings with it the same problems that face local communities in other countries. The very elements that make the culture interesting for the visitor are those that help change it. The Kensington Market in Toronto is a place that has become one of the most clearly identifiable environments in the city. People are attracted to its busy and densely populated streets particularly on Saturdays when the market atmosphere is one of colorful confusion—people and cars, fruit stalls and clothing bazaars jam the streets; caged chickens and ducks wait to be slaughtered; dirt, noise, and aromas, good and bad, fill the air.

Yet the introduction of well-to-do, Western middle-class values to a place created by a different culture has contributed to its deterioration as a special place. When Saturday shoppers started objecting to the unsanitary conditions and market litter on the streets, the caged birds suffering in the hot sun, the killing of poultry out in the open, the mini livestock farms in the backyards, they were complaining about the very things that lay at the heart of the area's special identity and that attracted people there to begin with. In subsequent political action, by-laws were enacted preventing people from keeping or killing livestock in the city. In other words, one of tourism's dilemmas is found in the urge to make people conform. And conformity contradicts the inherent diversity of urban vernacular forms and culture.

Environmental Values and Sustainable Futures

A new vision of the environment by the tourist industry has suggested that the use and preservation of natural areas and scenic places presents not conflict but opportunity for a new, harmonious relationship between man and nature. Rather than exploiting natural resources, tourism focuses on conservation as the basis for growth.[23] Armed with both a significant disposable income and a good education, tomorrow's tourist will be highly sophisticated and demanding of excellence in both facilities and service. The age of re-integration, it is said, will increasingly see man at one with the environment.[24] Impeding the realization of such a goal, however, is the very nature of resource-based industries, of which tourism is one. Several issues need to be considered.

Consumption versus Preservation. Even though there is a growing emphasis on the protection of natural areas, tourism is largely based on the consumption rather than the preservation of nature. For instance, in a 1987 White Paper on the future of tourism in northern Ontario by the Northern Ontario Tourist Outfitters Association, an agenda was outlined for the enhancement of the tourist industry. Among the highlights of the White Paper were strategies and recommendations for attracting the American traveler to the province's opportunities for moose and bear hunting and fishing.[25]

Free Resources. Leisure and travel inherently involve no lasting commitment to the ecological integrity and sustainability of the places people visit. Travel from one's home to other places for short-term leisure tends toward behavior patterns that are self-centered, exploitive and un-involving. This is particularly true of places that are alien to one's own culture. It is also true of landscapes that come "free": the natural scenery of mountains, lakes, and forests, commonly known as "scenic resources." The tourist must have a high level of intellectual and ethical understanding to become a champion of the environment and its protection. Up to now, history has shown that environmental ethics, in the abstract sense of promoting the preservation of nature, has not led to a better environment, or to the maintenance of regional diversity.

JOHN LIVINGSTON has written of the wildlife conservation movement that "wildlife preservation is a catastrophic, heartbreaking disaster."[26] In spite of the efforts of conservationists the world over, wildlife species continue to disappear from the face of the earth. It is true that there have been some successes: the whooping crane survives thanks largely to intense media coverage over many years. The white pelican has staged a comeback, ironically saved by an insurance company that uses the pelican as its corporate symbol.[27] But seen as a whole these events are mere blips on a continually deteriorating environmental future. The

root of the problem for the dilemma that Livingston describes has to do with society's difficulty in entertaining tangible feelings of responsibility for our descendants, and the values associated with the industrial growth economy as an article of faith.[28] The very term *scenic resource* described by the tourist industry and environmental planners implies use. If it can be seen as having direct benefit to society, it has value; otherwise it can be discarded as having none.

An environmental ethic, therefore, is central to efforts to protect nature and raise public awareness of these issues, a fact demonstrated by the existence of countless conservation organizations. But the history of the vernacular shows us that changes in the way things are done occur when it is perceived to be clearly necessary and practical to do so. Part of the environmental problem of tourism lies, as I alluded to earlier, in the absence of an emotional investment in the place. When this can be achieved, however, the opportunity for the collaboration of tourism and nature becomes possible. An example is the protection of coral reefs by tourism operators such as Club Med, where private ownership in many ways provides the best protection against destruction. Although these enclave resorts are not intended to benefit local people and could, by ignoring local culture, be seen as another symbol of internationalism, they do protect highly sensitive ecosystems.

Some countries, such as Tanzania, encourage enclaves, not out of desperation to attract tourism, but rather to try to avoid some of its negative cultural impact.[29]

The Special versus the Ordinary. Tourism looks for the special, the spectacular, the unusual. It pays little heed to an understanding of what lies beneath the facade: the natural forces shaping the breathtaking vista, the vernacular forces that shaped the picturesque old towns, or those shaping today's environment. In this context, the ordinary, representative, and familiar landscapes that we know and live in have to be understood in the philosophical context of integrated natural and human processes. The issue also has to do with pressure from sheer numbers of people and the damage they do to special places. For example, the thousands of enthusiastic birdwatchers who yearly visit Point Pelee National Park on Lake Erie for the spring and fall bird migrations trample sensitive flora into the ground and erode sand dunes. In the 1970s, pressures on its six-square-mile (fifteen-square-kilometer) area amounted to 15,000 people per square mile—an unsustainable impact on a small, sensitive site.[30]

THE INSPIRATION for environmentally viable tourism, then, must be based on a number of themes.

Education. J. B. Jackson has observed that what seems to be overlooked in the problems of tourism is its long and valuable educational tradition and the contribution it has made, both to the discovery of the world and to our way of interpreting it.[31] Conservation must begin by focusing on the emotional investment and commitment to the places we take for granted, after which it then can reach into the environments we visit.

The Cost of Doing Business. The protection of regional diversity lies in the perceived necessity for the tourist industry to pursue conserver and environmentally sustainable goals in the interests of making a living off the land. Paul Wilkinson has suggested that in most other forms of development some environmental values have to be sacrificed in return for expected human benefits.[32] But since tourism is basically dependent on unspoiled environments, the maintenance of these values at the highest level is essential. Protection of environmental quality becomes an internality—part of the cost of doing business.

Conservation Ethics and Spiritual Values. The establishment of a national park in the Queen Charlotte Islands of British Columbia in 1987 has shown how a nationwide conservation ethic can be a powerful force in protecting wilderness in places far from home. The *political* arguments for their protection were based on their value for tourism and recreation. Untouched forests and wild places will provide more income in perpetuity than the short-term profits derived from logging or mining. The Queen Charlotte Island coastal forests are as spectacularly beautiful as any on the Pacific coast and are regarded as a unique ecological habitat. They are among the most endangered virgin forests in the world and are the ancestral home of the Haida Indians. They represent the essence of British Columbia, the symbol of its verdant and wild landscape. Opposed to the preservation of this region was a powerful logging lobby dedicated to cutting any trees that can be taken for a profit. When efforts to prevent logging of the South Moresby area were overruled in 1974 by the provincial forestry service, it was the Haida who initiated a national debate by blocking a logging road and being arrested, in the glare of the media, by the Royal Canadian Mounted Police. Support grew as publicity and the media focused attention on the threat to the forests, and in 1987 the federal and provincial governments signed an agreement to preserve the region.

What is significant about the event is that the environmental values that were applied were not economic but spiritual. David Suzuki points out that the relationship of the native people to the land comes from traditions that are radically different from

ours. Yet, in our terms, their views make ecological sense.[33] There are signs everywhere of an environmental ethic that, although piggie-backing on the economic imperatives of tourism, is nonetheless a force of growing significance. In the distant future it may be possible to conceive of a nation of people who are spiritually, rather than physically, linked to the land—where the human presence in a treasured place can be replaced by the knowledge that it is simply there. To some extent this already exists in remote national parks that are too far away or too difficult to reach; where one's knowledge of them is gained through television programs and books, and where the idea of a "park" is related to protecting nature rather than serving human use.

Control. The accepted tools of landscape planning involve controls that limit numbers of people and that manage access and human contact with special places. But the most crucial issues are those of scale and local community control over its future. Where small-scale tourism facilities (such as low-rise cottage development), energy efficiency, and limits on numbers of people are integrated into the planning of travel experience, the likelihood of a healthy environment is greatly enhanced. This presupposes the determination by local people to order their own priorities in the process of benefiting economically from tourism. It involves control over how and where development should occur and the community's views on the relative priorities between tourist development and the protection of their traditions.

Opportunities for Environmental Experience

True understanding and commitment to the environment begins with the emotional and the intellectual experience of ordinary events and phenomena one finds everyday in familiar but often ignored surroundings. Tourism can best contribute to the identity and health of places when it is founded on environmental learning about the places one visits. An emotional investment in one's own place may also be the stepping stone to an investment in someone else's. The following case study of a forest education program illustrates one way in which this may be achieved.

PETAWAWA NATIONAL
FORESTRY INSTITUTE
In 1972 the Canadian Forestry Service initiated a public education program at the Petawawa National Forestry Institute, a federal research station studying forests and forestry in Ontario. Its purpose was to develop a long-range plan for public awareness of forests based on the institute's land resources.[34] Its general direction and objectives were influ-

enced by a number of concerns: how forests work; how they should be used; the question of government jurisdiction; and the research functions of the institute itself. The institute uses over ninety-eight square kilometers of experimental forest land in the Ottawa Valley at the center of the Great Lakes–St. Lawrence forest region. Its research includes the study of various aspects of forest soils, ecology, tree genetics and tree growth, as well as silviculture, and forest management techniques and related activities.

The basic objective of the interpretive program was to give people a balanced view of issues, among them: forest ecology; the cycle of energy and nutrients; the interdependence of producers, consumers, and decomposers common to all ecosystems; the processes that give rise to the visual landscape; the concept of change as a fundamental part of the dynamics of nature; the presence of human influences, such as forestry and recreation, on the landscape. It was also important to establish connections between various systems: the similarities between forest and aquatic ecosystems; the dependence for water supply in cities on forests and watersheds and their management; the connections between coniferous forest regeneration and fire.

People need a basis for making informed judgment on the many issues that are the focus for public debate. The conflicts that arise between professional forestry groups and conservationists have led to extreme views where the basis of the message is determined by who is trying to educate whom. The one-sided view is, for instance, the downfall of many parks programs that are geared to stress natural history and the beauty of nature, but that omit the more unpleasant realities of overcrowded campsites and polluted park environments. Similarly, foresters who promote the universal benefits of forestry and underplay its environmental effects ignore the realities of its destructive impacts.

Two factors were of particular significance to the program. First, the site was typical of much of the Canadian shield landscape; therefore its representative nature helped determine the kind of program that could be offered. Interpretation of the landscape and human activities on the land depend on what is there. This landscape is *not* remarkable in any way. There are few, if any, unique natural features or spectacular falls or views. Under these conditions, appreciation and interpretation of the site becomes a process of becoming aware of a familiar but largely unnoticed or misunderstood environment. The object of the public awareness program, therefore, was to give new meaning and significance to the objects, activities, and places that are part of the daily surroundings of the summer traveler.

Second, to be understood by visitors, the site resources had to be seen as part of a process—a set of causal relationships between nature and people. The story of the regeneration of an old field, for instance, could not be told without an understanding of the ecological processes that give rise to regeneration, the logging and agricultural history that preceded it, and the management processes that insure rehabilitation of a disturbed landscape. Such considerations, determined by the site, provided a basis for planning.

The central theme of the program was "Forests and People," in which the visitor could find answers to some basic questions about forests. How do they work as natural systems? What is their value to people? How will they be sustained in the future? The conceptual program was thus developed to explore the three interconnected ideas of ecology, use, and management. From this central theme, subthemes were added to explain in greater depth various aspects of the forest. They included ecology, commercial timber, wildlife, historic land use, codes of behavior and land ethics, and research.

The approach to telling the story was based first on the aims of the program (objectivity, process, connections), and second, on the inherent opportunities the site provided. In the main theme and more detailed subthemes, links were established between the various components of the subject. The intent was to main-

Dramatizing a hillside pine forest.

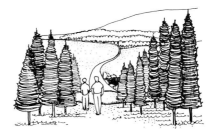

The unexpected view.

**Principles for
site planning
and interpretive
methods.**

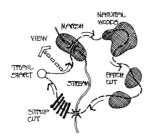

*The importance of scale in maintaining
interest.*

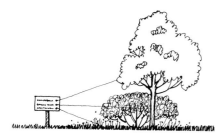

*Economy of means: getting the most results
from minimum effort and resources.*

*Involvement is a key element in
environmental education.*

*Appropriate materials are those that fit the
landscape and are easy to work with.*

The forest trail interprets the difference between two forest ecosystems. Surprise and involvement enhance understanding.

tain the idea of causality and of continuing cycles between nature and man. For instance, in the ecology story the message would reveal the nature of the site, nutrient flows, water regime, types of vegetation responding to the site, wildlife supported by the vegetation, the historic and current uses of the forest, fire as an agent of pine regeneration and as a management tool, silvicultural techniques and forest products. This approach would demonstrate to people with specific interests how their interests are relevant and connected to the total picture of the forest and people.

The following principles for site planning and interpretive methods formed the basis for implementation.

Exploiting the Site. The primary purpose of this principle is to give meaning and significance to a familiar but largely unnoticed forest landscape. Sensitive site design involves revealing the inherent quality of the site— its genius loci. It is central to the visitor's enjoyment and consequently his understanding of the landscape. Through site design the contrasts among forest environments are emphasized and a representative landscape becomes memorable.

Scale. Impressions of the landscape vary with the mode of travel. To someone walking, the range of interpretive elements must vary as much as possible because the same scenery quickly becomes tedious. Rapidly contrasting experiences are therefore required to maintain interest and understanding. Thus scale becomes a basis for trail alignment.

Involvement. The well-known dictum that the more a person does for himself the greater his learning experience also implies that the simpler and more direct the relationship among the visitor, the story to be told, and the site, the more effective will be his experience.

Economy of Means. Budget is always a crucial limitation in program development, even when the importance of an environmentally enlightened public is generally acknowledged. Thus simplicity and ingenuity must predominate.

Appropriate Materials. The materials used for floors, trails, benches, and picnic areas must be convenient, wear well, and be appropriate to the urban visitor. They must also maintain a sense of the place and involve some effort on the part of the visitor. Acres of mown grass and smooth paths, typical of some facilities, detract from the experience of the forest landscape as a place and smother its essential character. In design terms, the ability of materials to protect the environment through which the visitor is passing is essential.

THE MOST important aspect of environmental understanding is the need for an informed, unsentimental perspective on nature that will link land, society, and conservation and that will provide an environmental basis for tourist values. The preceding example and others like it are attempts to provide a foundation for understanding familiar rural landscapes, but the problem of tourism and education must also be seen in the context of urban settlement. It is worth reemphasizing that the economic survival of many a beautiful and historically rich town depends on the wealth that tourism brings, but that tourism in turn also provides the prescription for the town's eventual demise. When faced with the need for economic recovery, few depressed towns are able to resist the lure. Tourism, in fact, has become the universal solution to a more prosperous future. The following case study shows how a community in the process of change has benefited from tourism while preserving its intrinsic values and identity.[35]

MANTEO, NORTH CAROLINA
The story is a familiar one; only the details change. A once thriving town is bypassed by a new highway built to bring traffic direct to holiday ocean beaches. Visitors become fewer, small industries and shops close, unemployment skyrockets, and a way of life is threatened. This is what happened to Manteo, located on North Carolina's Outer Banks on Roanoke Island. Best known as the site of Sir Walter Raleigh's lost colony of 1585, for its experimental freed-slave

colony during the Civil War and for its boat-building industry, Manteo became important as the Dare County seat because of its central location and protected harbor.[36] Early in the twentieth-century a bridge linking the Outer Banks to the rest of North Carolina and Virginia made the previously isolated Roanoke Island accessible to visitors. Tourism replaced boat building and fishing as the primary economic activities. Manteo was eight kilometers from the beach and, not being situated directly on the highway, was consequently bypassed by the tourist traffic. Its prosperity plummeted; waterfront activity died as downtown businesses moved to the beach or the highway strip; seasonal unemployment rose and land values sank leaving underused buildings, land, and infrastructure. Yet its smallness, intimacy, natural beauty, village character, and rural past were attributes that could be exploited for tourism.

The danger of taking this route to economic recovery, of course, has led to the demise of many communities as their traditions and important places fall victim to the phony folk culture and frozen wax museum authenticity of the "Preserved Historic Town." Manteo's recovery, however, was the result of a remarkable process of grass-roots community leadership and development guided by a planning process that helped the residents identify and preserve what they valued about life and about their

landscape in the face of change. Randolph Hester comments that these important social patterns and places came to be called the "Sacred Structure" by locals and inspired a plan for community revitalization and development that was controlled by them.[37] Questionnaires and interviews with the townspeople about their attitudes toward their town showed that while they wanted both economic regeneration and development they also wanted to retain the small town atmosphere. This latter wish meant that something lay hidden beneath the surface, an unexpressed attachment to the place—the style of living and landscape—that was essential to its cultural continuity.

Planning focused on behavior mapping that recorded what people did and where they did it—things that were not revealed in the standard surveys. Activities like the exchange of small talk at the post office, hanging out at the docks, checking out the water for the tides, the fishing, and the weather, happened in the same places every day. Daily rituals indicated a dependence on specific places that could be disrupted by changes in land use. A list of these was developed, and people were asked to rank them in order of their significance, and to indicate which ones could be sacrificed in the interest of tourist facilities. From these was published a map of places that people wanted protected from future

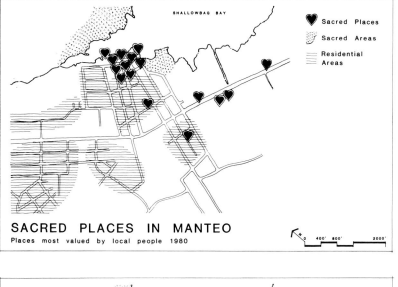

SACRED PLACES IN MANTEO

Places most valued by local people 1980

Legend:
- ♥ Sacred Places
- ⠿ Sacred Areas
- ≡ Residential Areas

SHALLOWBAG BAY

N 0 400' 800' 2000'

Manteo
North Carolina
Planning Surveys
1980.
(Randolph Hester)

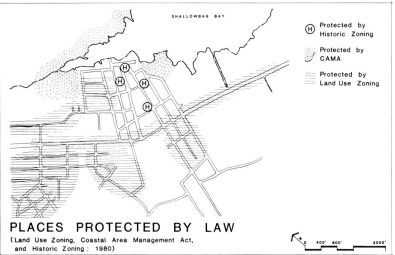

PLACES PROTECTED BY LAW

(Land Use Zoning, Coastal Area Management Act,
and Historic Zoning : 1980)

Legend:
- Ⓗ Protected by Historic Zoning
- ⠿ Protected by CAMA
- ≡ Protected by Land Use Zoning

SHALLOWBAG BAY

N 0 400' 800' 2000'

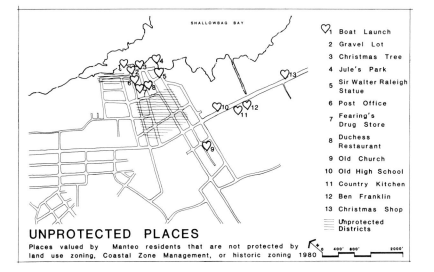

UNPROTECTED PLACES

Places valued by Manteo residents that are not protected by
land use zoning, Coastal Zone Management, or historic zoning 1980

Legend:
- ♡1 Boat Launch
- 2 Gravel Lot
- 3 Christmas Tree
- 4 Jule's Park
- 5 Sir Walter Raleigh Statue
- 6 Post Office
- 7 Fearing's Drug Store
- 8 Duchess Restaurant
- 9 Old Church
- 10 Old High School
- 11 Country Kitchen
- 12 Ben Franklin
- 13 Christmas Shop
- ≡ Unprotected Districts

SHALLOWBAG BAY

N 0 400' 800' 2000'

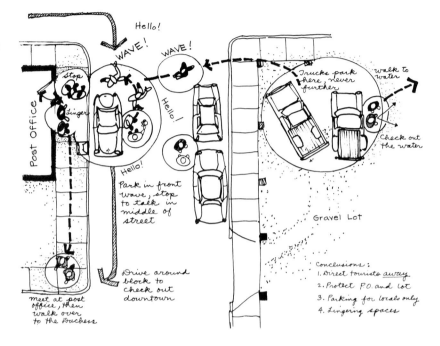

Newsing at the Post Office. (Randolph Hester)

development. Such things as the marshes surrounding the town, the drug store, the post office, a statue, the town hall, locally made street signs, the cemetery, and the Christmas tree in the gravel parking lot became for the local people the "Sacred Places," for which they were prepared to sacrifice economic gain. They embodied, in fact, the very life of the town.

The "Sacred Structure" map eventually became part of the planning board's formal discussion on future planning for the town. As Hester has described it, it influenced the plan in a number of ways. It focused normally vague discussion on values

into specific decisions about which places should be preserved and which places could be changed to accommodate tourists. Interestingly, other town documents, such as zoning ordinances, lists of historic sites, and traditional inventories of significant architecture did not include the places essential to the social life of the town.

The "Sacred Structure" inventory was used by residents to evaluate objectively how development plans would affect the places that were important to them. It directed the choice of the final plan. Of seven options that were presented to the community the one chosen was less

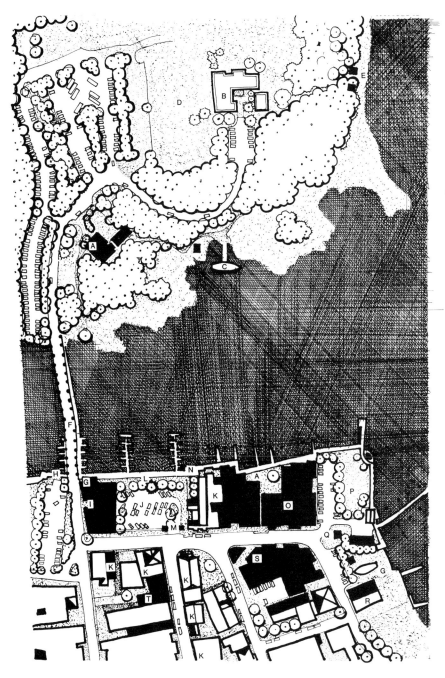

*The Manteo
Village Plan.
(Randolph Hester)*

MANTEO VILLAGE PLAN

A	Living-Learning Center	K	Sacred Buildings
B	Performing Arts Center	L	Sacred Tree
C	Elizabethan Ship	M	Sacred Gate
D	Anglo-American Folk Festival	N	Boardwalk
E	Beach	O	New Housing Commercial Condominiums
F	Bridge	P	Jule's Park
G	Exhibits	Q	Sacred Statue
H	Town Launch	R	Boat Building Center
I	Inn	S	New Housing-Commercial
J	Sacred Lot	T	Infill Buildings

Manteo's recovery involved local people making the place their own and controlling change. Here, an outdoor art show. (Randolph Hester)

economically advantageous but respected more of the places people valued most.

It provided a basis for negotiation with developers. Inappropriately scaled, comprehensive urban redevelopment proposals were precluded in favor of small developments that maintained the integrity of essential places. Performance criteria were provided to insure that development would be consistent with the preservation of important features such as the marsh; with infill building rather than by wholesale destruction and reconstruction of the town's existing built fabric.

It raised valued community places to the same level as economic, technical, and political considerations. This permitted the town to benefit from tourism while protecting important places from tourist encroachment, in effect insuring that the life of the community could retain its viability. In order to reduce the legal vulnerability of the plan a number of mechanisms were used that included state historic corridor legislation, formal adoption of the guide to development, and the creation of a village business district requiring a conditional use permit for all new development. The guide required that any new development or change in the village must be assessed by local people for its impact on the use of the "Sacred Structure": for instance, it should protect or enhance local pedestrian access, parking, and local visual access.

*Outdoor cafe
on the quay.
(Randolph Hester)*

*Construction of
a new wharf.
(Randolph Hester)*

THIS EXPERIMENT in socially based design is significant in a number of ways. First, it deals with places in terms of their importance to the people who use them, and it reflects in a real way the need to plan for the experiential qualities advocated by Lynch.[38] The visual significance of these qualities cannot be measured according to the conventional standards of traditional planning criteria. The Christmas tree in the gravel parking lot, for instance, had significance for the people of the town, but none for outsiders. Consequently, it would not have been mapped as having visual importance. Among Manteo's significant places only two were protected by historic preservation legislation or zoning laws. Like the places where local children like to meet and play, their importance lies in subconscious needs, not in their visual attraction, and they are the product of essentially vernacular social forces. Second, the process illustrates one mechanism for how communities, faced with the rapid and disruptive change that tourism brings, can maintain their valued social structure and places while enjoying its economic benefits. Third, it illustrates the potential for beneficial change through community action, where people who have made a place special continue to be instrumental in its development rather than falling victim to change that only benefits others. It illustrates, too, that the maintenance and protection of places is fundamentally linked to the social and physical forces from which they evolve, and that the static preservation of buildings, gardens, towns, and historic landscapes must take second place to the real issue of preservation—the maintenance of continuity.

8 / Principles for Regional Design

What role does design play in the development of a contemporary regional landscape? A historical perspective suggests that the differences between one place and another have arisen, not from efforts to create long-range visions and grand designs, but from vernacular responses to the practical problems of everyday life. Indeed, it can be argued that purposeful design has done more to generate placelessness than to promote a sense of place. The new forces shaping the landscape are no longer small and local in scope but are great in scale and consequence. The technological and economic impact of these forces on the environment has never before had such profound potential for the destruction of life systems. As a discipline dedicated to fitting man to the land and to giving it form, contemporary design is faced with solving problems that have traditionally not been a part of the agenda in the creation of vernacular places.

In the past, there were limits to what one was able to do and the extent to which one could modify the natural environment. The constraints of environment and society created an undisputed sense of being rooted to the place, but they were, nonetheless, limitations to be overcome, not inherent motivations to be at one with nature. In today's landscape the heterogeneity of the past is giving way to a more homogeneous, information-based society. In design terms, therefore, it becomes as much romantic nonsense to force the old regional differences upon this new landscape as it is to expect people to give up cars, washing machines, and television in the interests of a better environment. We are locked into our times and ways of doing things.

Yet, there is a dilemma for designers in the new and evolving landscape. The determinants that shaped the settlements and countryside of preindustrial society and that gave rise to the physical forms which we now admire are now no longer those of environmental limitation but of choice. Creating a sense of place involves a conscious decision to do so. At the same time, the need to invest in the protection of nature has never been so urgent. The connections between regional identity and the sustainability of the land are essential and fundamental. A valid design philosophy, therefore, is tied to ecological values and principles; to the notions of environmental and social health; to the essential bond of people to nature, and to the biological sustainability of life itself. This is the new necessity that will counterbalance and bring some sanity to a world whose goals are focused on helping us "live in a society of abundance and leisure."[1] Yet values that espouse a truly sustainable future will only emerge when it is perceived that there are no alternatives. It is possible that over time the fragility of earth's life systems will create an imperative for survival on which a new ethic can flourish. The international agreement to protect the earth's ozone layer, signed in 1987, may be one indication of this trend. And it is only on this basis that regionalism can become an imperative—a fundamental platform for understanding and shaping the future landscape.

IN THE preceding chapters I examined the various factors affecting regional identity in order to establish a framework for a design philosophy for the contemporary landscape. This chapter suggests the principles that seem most appropriate to this objective.

Knowing the Place

Recognizing how people use different places to fulfill the practical needs of living is one of the building blocks on which a distinctive sense of place can be enhanced in the urban landscape. Regional identity is connected with the peculiar characteristics of a location that tell us something about its physical and social environment. It is what a place has when it somehow belongs to its location and nowhere else. It has to do, therefore, with two fundamental criteria: first, with the natural processes of the region or locality—what nature has put there; second, with social processes—what people have put there. It has to do with the way people adapt to their living environment; how they change it to suit their needs in the process of living; how they make it their own. In effect, regional identity is the collective reaction of people to the environment over time.

At the turn of the century Patrick Geddes taught that before attempting to change a place, one must seek out its essential character on foot in order to understand its patterns of movement, its social dynamics, history and traditions, its environmental possibilities. He commented on the way planners dictated form and solutions to problems with little reference to the reality. In his design studies for Madura in the Madras Presidency he wrote, "One of the poor quarters . . . is at present threatened with 'relief from congestion' and we are shown a rough plan in which the usual gridiron of new thoroughfares is hacked through its old-world village life. . . . the sanitary improvements begin by destroying an excellent house for the sole purpose of inclining the present lane from the position slightly oblique to the edge of the drawing board to one strictly parallel to it."[2] In effect, he was saying that modifications to city plans, and for that matter modifications to any landscape, are based on thought processes that begin and end with paper, not the environmental and social realities of the place.

Underlying every urban or urbanizing environment that has developed an image of increasing sameness are unique natural or cultural attributes waiting to be revealed. A place's identity is rarely completely destroyed. There are always elements of the original landscape that remain, sometimes deeply buried beneath the new. Landform, remnant native plant communities, an old hedge, a barn, old paving stones speak to natural and cultural origins and changing uses. The task is to build an identity based on these remnants.

The hidden elements of a place affect our senses, albeit unconsciously. Tony Hiss describes this in his analysis of experiencing places: "Small, unnoticed changes in level play a larger organizing role in our activities than we suspect: in Manhattan, the right-angle street grid, which keeps people's eyes focused straight ahead, and the uniform paving of streets and sidewalks, together with the solid blocks of buildings on both sides, tend to keep New Yorkers from noticing the natural contours—or what's left of the natural contours—beneath their feet. The nineteenth-century Manhattan developers who covered midtown fields and meadows with brownstones did such a good job of lopping off the tops of hills and filling in valleys that a hundred years or so later . . . no one really knows what the original topography was. . . . Nevertheless, almost every block has some rise or dip to it, and these hints of elevation do help people define certain districts."[3] Several other examples of natural and cultural attributes will illustrate

how these affect our sense of a place's identity.

Identity through the Landscape. The deep, densely wooded ravines of Toronto that were cut from tableland by streams following the last ice age are part of a major system of rivers draining south to Lake Ontario. Twenty-one meters below the flat, urbanizing plateau of the growing city, they formed a unique system of remnant southern hardwood forest and streams, a habitat for animals and birds within an urban area, a place where the original forest and the natural history of the land could still be experienced. It was where sounds of traffic were no longer heard, where smells and tactile feelings were enhanced by the utter contrast of enclosing woods, suddenly experienced as one reached the valley floor from the level of the street. As the city expanded, many ravines were obliterated by encroaching development, and by the 1950s, 840 acres (340 hectares) of the 1,900 acres (770 hectares) of original ravine land in the city had been given over to houses, factories, and roads.[4] The unique character of Toronto's landscape—the city's structure and identity—rapidly gave way to featureless urban growth. Others were left alone, not as a consequence of planning but because they were simply a nuisance, or difficult, to fill in. Today they are recognized for their significance to the environmental and social well-

Knowing the place. Toronto's ravines are its most important linear natural feature, giving the city an unmistakable stamp of identity.

being of the city. Within the urbanized environment they have become one of the key elements that make Toronto different from other places, both as identifiable landscape form and for the uniquely adapted recreational activity they have generated. Protected by dense woodland below the level of the city that eliminates winter winds and provides a markedly cool summer climate, they are ideal places for winter skiing, walking, birdwatching, and nature study. They have become Toronto's stamp of individuality.

Different Places for Different People. Many of our urban parks have developed as cookie-cutter patterns of grass and trees, models imposed on the city by a tradition of standard landscapes for standard people. These values have dominated design thinking and helped create a landscape that does little to reflect the inherent social diversity of neighborhoods. The quality of urban life today has to do, among other things, with the recognition that diverse social groups need diverse landscapes, that choices between one place and another must be available. A city that has places for foxes and urban woodland, regenerating fields and urban wilderness, is more interesting and pleasant to live in than those that lack such places. The identity of the urban landscape is also based on "hard" urban spaces: busy plazas and markets; noisy and quiet places; culti-

vated landscapes and formal gardens; funfairs and cultural events. In multicultural cities, there are also many social needs that should be addressed by public spaces that reflect the cultural and physical identity of these groups. The elderly—an increasingly large group in North American cities—garden, cycle, bowl, take photographs, play cards and board games, watch the world go by. Ethnic groups, many of whom are moving out to the suburbs of large cities, take with them traditions of productive urban farming, festivals, family gatherings for picnics and intensive use of small spaces that are transforming hitherto conventionally landscaped suburban places. Over time, changing community structure in many suburban areas will create a new kind of environment, one that fits the people who live and work there. Thus by-laws and restrictions on what may or may not be done on private property, which have previously dominated the social and physical character of many suburbs, will need to be modified to reflect new communities values.

The socially disadvantaged, the bag ladies, the homeless, and transients use parks and waste places in the city night and day and often year round. As a group, these people use the established parks more permanently and with more basic need than the people for whom the parks were actually intended. Yet these are the people who are not welcome in the

Creating different places for
different people is the building
block for establishing natural and
cultural variety, and thus a sense of
place. Cities need:

*Places for photographers and gardeners.
(Parks and Property Department,
Municipality of Metropolitan Toronto)*

*Wilder places for people seeking quiet and a
natural setting. (Suzanne Barrett)*

*Places for older
adults. (Suzanne
Barrett)*

Places for the disadvantaged who use many public spaces more often and more consistently than the people for whom they are intended. (Chris Winter)

Safe streets for children as well as for cars. Local streets are often the best public spaces. (Suzanne Barrett)

parks system. The use of streets as community spaces by adults and children in healthy neighborhoods establishes an essential vitality, social character, and commitment to place that comes from common use. They are the spontaneous result of people using the street as a natural meeting place that the standard park cannot fulfill.

Randy Hester has shown how the daily ritual of people exchanging gossip, meeting friends, and negotiating business is place specific. The places that have significance for people, such as the local post office, corner store, community park, or parking lot, are those that "have become so essential to the lives of the residents through use or symbolism that the community collectively identifies with these places."[5] Most of these may have no appeal to the designer's eye as being "beautiful" or worthy of preservation, but they are nonetheless a basis for healthy communities.[6]

Maintaining a Sense of History

Rarely does the designer have the luxury, or more appropriately, the misfortune, to create a place from scratch. Something is always there before he begins: a history, a peculiar character, a meeting place. Design inevitably involves building on what's there in the process of change.

The protection of natural and cultural history—the reuse and integra-

tion of the old into the new without fanfare while avoiding the temptation to turn everything into a museum because it is old—lies at the heart of maintaining a continuing link with the past and with a place's identity. Our overwhelming desire to eliminate our past is nowhere more evident than in the destruction of nature that we find in every corner of the globe in large or small measure. Similarly, the tendency for the new in urban development to destroy the old in the interests of economics is one of the major reasons for placelessness in the changing urban landscape. There are no longer any historical reference points by which one understands where one has come from in the process of building the new. The remnant native plant communities that still survive in protected parts of the city—in cemeteries, valley lands, older residential areas—also link us with the past, with the pre-development landscape, and with the historic interactions of man and environment. Evolving and fortuitous naturalizing plant communities in the city's forgotten places—railway corridors, abandoned lands, industrial properties, the corner spaces found in every city lot tell us more about the dynamics of natural processes and the sustainability of nature in urban areas than those that have been imposed, in aesthetic, or horticultural terms, on the environment.

The same is true of new urban developments that have ignored cultural history. The absence of fine or significant architecture has often been a red herring in arguments about the worthiness of preserving and reusing old buildings. The basic purpose of maintaining old parts of town is to link us with the past—to enhance one's knowledge of a place's cultural roots. An example is the redevelopment of urban waterfronts that since the 1970s has become one of the most important development trends in North America.

Rivers flowing through the industrial cities of eastern Canada have evolved as essentially urban places like Quebec City.

The way in which cities have developed and have been modified by their waterways has varied considerably from place to place. This is particularly apparent in countries of intense regional differences such as Canada. In the western prairies, for instance, the flatness of the land and great distances from the mountains to the sea have produced shallow, winding, erosion, and flood-prone river landscapes that were unsuitable as major modes of transportation. Patterns of urban development evolved in response to these environments. The rivers, essential as a water supply, have nonetheless been historically neglected, or used as convenient dumping grounds or for water-dependent industry. They were not integrated with the cities that grew around them until their potential for recreational use was realized. One exception is the city of Saskatoon, Saskatchewan, which has had a long tradition of preserving open space. The temperance colo-

Rivers in the western prairies produce shallow, winding, and flood-prone landscapes. They have evolved into the "green valley" open-space systems of cities like Saskatoon, Saskatchewan. (Meewasin Valley Authority)

nizers who settled this prairie city in 1884 dedicated the river banks as public land and set the pattern of connected linear open spaces along the river for the future.[7] Today, western rivers project the distinct and immediate image of a prairie link between city and farmland. Their identity as places lies in their intrinsic "naturalness" in the way they have evolved into "green valleys" that wind through urban environments seemingly keeping the fabric of streets and buildings at bay. In contrast, the industrial cities of eastern Canada, such as Montreal or Toronto, which grew on the shores of the deep, wide-flowing rivers and waterways of the Great Lakes system, have provided the essential water transportation from which has grown the commercial and industrial base of Canada.

It is the environment of work—the vast scale and drama of large cargo boats and cranes, quays, railways, and storage sheds—that makes these waterfronts so exciting and gives them a special presence. They symbolize the oceans, travel to strange and exotic lands, and international trade. All great waterfronts have this quality; Istanbul, Halifax, and Stockholm are among them. For many cities, however, the old functions of the working port have given way to recreational ones. Redevelopment, too, frequently leads to the total destruction of the previous landscape of industry, grain silos, and railways that used to provide their economic base and their reason for being. The observer of many new developments may be excused the temptation of wondering why the previous landscape of industry, often carefully preserved in old photographs, looks so much more interesting than the new commercial developments that have replaced it. In 1987 the city of Toronto rejected a proposal to retain the old grain silos on its waterfront as historic landmarks. A letter to the local newspaper on this issue made the following point: "These silos are a vital part of the history of Canada in the same way as the brooding castles of Europe reflect the past. Both structures were born out of the needs of their time and have become symbols of an era and a country. . . . we would today take a rather dim view of a council in Europe that had condemned their castles to demolition because they had become useless. The silos are not useless in the memory of a collective Canadian consciousness."[8] The making of memorable places is linked to history.

Environmental Learning and Direct Experience

Environmental literacy lies at the heart of understanding the places with which we are familiar, and thus at the heart of the issue of identity. It is necessary for people who live in and use urban places, indeed places

Toronto's changing waterfront expresses a sense of history and a place of work.

of any kind, to know the environment around them. An awareness of place can only be enhanced when it becomes a part of people's everyday lives. Formal school programs, like the once-a-year visit to the country to "educate" urban children in nature lore, do little to engender or deepen knowledge of the environment, or more importantly, to encourage environmental values. These are more likely to come from understanding the places that are close to home. The same principle applies to the interpretive programs provided for the enlightenment of adult campers in provincial parks, that explain the workings of unspoiled nature out in the woods, but totally ignore the problems of water pollution, deterio-

ration of vegetation, garbage dumps, and disruption to wildlife from human presence that occurs in the campgrounds themselves.

An urban waterfront on the Great Lakes, for example, can speak to its place within the system. Through its entertainment and cultural facilities, it can capitalize on key subjects and issues of the region, such as the international implications of pollution that begin here, its history of sailing vessels, trade, and discovery, the aquatic ecosystem, local plant and animal communities, and the interrelationships between people and environment. But the marine aquaria, tropical fish tanks, captive porpoises, and killer whales that leap through hoops and kiss the girl in the bikini for our entertainment that have become the prime attractions of many waterfronts tell us nothing about the place and are environmentally and ethically bankrupt. They contribute instead to environmental ignorance, to a lack of context and identity. It is possible, though, to reinforce a sense of place through educational exhibits that are at once instructive and fun. An example, was the headquarters building of the now-defunct Greater London Council. Two large fish tanks lined the entrance foyer to the building. One showed what the Thames used to be like in the days of the river's worst pollution—a lifeless murky underwater environment. The other showed the rehabilitated Thames as it is today with the dozens of species of fish that now live in its much improved waters. It told one a great deal about the river on which London depends and about some of the environmental concerns of government at the time.

Knowledge through education of a place's environmental or cultural significance changes our attitudes and the way we experience it. Public reaction to a highway "no-mow" experimental program in North Dakota, for instance, was initially negative. In a survey of motorists about the program along the right-of-way, 82 percent of those interviewed said that if they had to make a choice, they preferred the mown plots to those that had been left unmown. However, when they were informed that the

unmown plots provided waterfowl nesting habitats, many wished to change their answer.[9]

Giving meaning and significance to ordinary and largely unnoticed places, whether this happens to be a suburban street, a few square feet of prairie, or a representative forest landscape is the basis of regional identity. The task of design is to encourage an understanding and enjoyment of the landscape that comes from both emotional experience and scientific knowledge. In this way, normally overlooked landscapes can become memorable.

Doing as Little as Possible

Kevin Lynch remarks, "A hunger for the control of large-scale form is all the more dangerous because it coincides with strong contemporary trends towards large-scale investment."[10] The pressures (that come from educational conditioning) to do as much as possible in making changes to places often appears endemic to the land design disciplines. In the absence of a basic ecological foundation on which design can rest, this is to be expected. Doing as little as possible, or economy of means, involves the idea that from minimum resources and energy, maximum environmental and social benefits are available. The greatest diversity and identity in a place, whether a regenerating field or urban wetland, or a cohesive neighborhood community, often comes from minimum, not maximum interference. This does not mean that planning and design are irrelevant or unnecessary to a world that if left alone would take care of itself. It implies, rather, that change can be brought about by giving direction, by capitalizing on the opportunities that site or social trends reveal, or by setting a framework from which people can create their own social and physical environments and where landscapes can flourish with health, diversity, and beauty.

Urban street systems, for instance, provide the overall physical framework within which neighborhoods flourish and diversify. Local by-laws and design requirements may enhance or inhibit the social and physical complexity of a community. In chapter 3 I discussed those situations where political power seeks to impose control on nature or humanity, thereby obliterating the inherent diversity of places. The over-regulation of what can be done to private property has an inherent potential to generate tedium. Compare the planned shopping arcades of many new developments, where regulations and design dictate the style and positioning of signs and setbacks, with the shopping streets that have grown up in response to the needs of individual store owners. The former somehow lack the vitality, life, and interest of the latter. Similarly, the formal landscaped avenues, parks, and gardens that grace the institutional centers of many cities and speak to their sense of civic pride lose their special identity as places when they become universal expressions of the city's landscape.

At another scale, it has long been an article of faith that the designers of public landscapes should be able to predict human behavior, on the basis, first, that behavior is indeed predictable and controllable; second, that it will not change; and third, that it is a necessary measure of a designer's competence. The isolated benches that no one sits on, the playgrounds that children avoid, the pathways and pedestrian routes that no one follows, and the gathering places without people bear witness to the emptiness of that claim. William Whyte's careful observations about what people in New York actually do in city spaces has demonstrated the key elements of design: understanding the psychology of behavior (how people actually behave, what they actually do) and how to bring in those elements that enhance the diverse use of public space.[11]

People need to control how they use the environment around them, and in the process of doing so the designed landscape becomes a vernacular one, responding to practical needs. As Whyte has shown, dynamic and interesting places can be created simply by locating a food-vending stand in a place where passersby can see it from the sidewalk, or by pro-

The principle of doing as little as possible may involve simply removing trash from a pond to enhance its beauty or environmental quality.

viding seats that can be moved around at will.[12] Similarly, the experience of a natural place can be enhanced beyond measure by uncovering a clogged stream so that its sound can be heard, or by removing trash from a pond so that its natural beauty is revealed.

In *City Form and Natural Process* I argued that the horticultural tradition has long been the basis for getting the least results for the most effort in money, energy, and manpower.[13] Yet it is not horticulture per se that is at the root of the problem. It is the lack of an ecological perspective that permits doctrine, or expediency, or both to impose similar environments on differing places. The desire for uni-

versal solutions is strong and lends credence to the adage that design style—those characteristics of a designer's work that identify him—are in fact a series of never questioned mistakes repeated over and over again. For instance, there is no doubt that the move toward natural landscapes is based on a genuine concern for greater variety and sustainability in our cities. Yet from the point of view of regional identity, the inspiration for naturalization can be tarred with the same brush as that which inspired our current "pedigree" urban landscape. It becomes another doctrine. The designer, determined to create alternative landscapes, finds himself tied to "wildflower seed mixes" that are, for the most part, drawn from plants found from the prairies to the east coast. The result is an international naturalized meadow replacing the international green carpet. Inspired by the native flora and roadside wildflower program in Texas, commercial seed mixes are often selected for the color and spectacle of their flowers. But they are often alien to the local region and are, consequently, taken over by native flora after a couple of years, which defeats the original, sustainable, low-cost objective. Only the local setting can create the kind of regional landscape that we are concerned with, and it is from here that one must draw inspiration. The natural communities that are indigenous or adapted to the place are those that occur with the least effort and with the greatest sustainability and variety.

As a design principle, doing as little as possible implies, first, an understanding of the processes that make things work; second, providing the structure that will encourage the development of diverse and relevant natural or social environments; third, knowing where to intervene to create the conditions for them to occur; and fourth, having the humility to let natural diversity evolve on its own where it will.

Sustainability

Sustainable landscapes are central to the regional imperative. Sustainability involves, among other things, the notion that human activity and technological systems can contribute to the health of the environments and natural systems from which they draw benefit. This involves a fundamental acceptance of investment in the productivity and diversity of natural systems. Conflicting points of view over the priorities of development versus the preservation of natural wealth have been the focus of discussion and argument for a very long time, particularly as it affects the Third World. The World Commission on Environment and Development, established by the United Nations in 1983, and whose report appeared in 1987, has examined and proposed ways in which economic development initiatives and environmental conservation might be reconciled. For this to be workable would require the development of an environmental ethic far different from current attitudes and perceptions that see nature as "resources for the benefit of mankind." Such a notion would seem practically to be unattainable. However, Maurice Strong, the Canadian member of the commission, has commented on the need for countries to shed their narrow concepts of self interest, parochialism and, in the economic field, protectionism.[14] Although he recognizes the odds against the emergence of such a world view, he sees no alternative: "The principle basis for optimism that the kind of changes I foresee as necessary will occur is, very simply, that they are necessary and therefore must occur."[15]

Irrespective of such a world view, however, the principle of investment in nature, where change and technological development are seen as positive forces to sustain and enhance the environment, must be the basis for an environmental design philosophy. Its principles of energy and nutrient flows, common to all ecosystems when applied to the design of the human environment, provide the only ethical and pragmatic alternative to the future health of the emerging regional landscape. And this leads naturally to the last principle.

Starting Where It's Easiest

This principle, borrowed from Jane Jacobs,[16] is fundamental to achieving anything in a world where the statistics of global environmental disaster are at once horrifying and numbing. Through the media, the visibility of environmental issues everywhere in the world is immediate, vivid, and emotionally involving. At the same time, these media reports have two things in common. First, they are almost inevitably out of town. They are somewhere else: in the diminishing

rain forests of Brazil; in the burgeoning population and desperate poverty of Africa; in the dying northern lakes of Canada and Sweden that are succumbing to acid rain generated by polluting industries a thousand miles away. We have the paradox that in a world increasingly concerned with deteriorating environments and explosive urban growth, there is a marked propensity to ignore the very places where most people live. Second, the issues are so enormously complicated and of such magnitude that most concerned people feel helpless to do much about them.

Beginning where it's easiest, therefore, has to do with where most people are and where one can be reasonably certain of a measure of success from efforts made, no matter how small. Successes in small things can be used to make connections to other larger and more significant ones. This is, consequently, an encouraging environmental principle to follow in bringing about change. It is, in fact, the only practical basis for doing so. In design terms, the regional imperative is about the need for environmental ideals that are firmly rooted in pragmatic reality. It is about focusing on things that work and that are achievable at any one point in time. It is about a concerned and environmentally literate community prepared to insure that the health and quality of the places where they live are made a reality; where the role of technology is integrated with people, urbanism, and nature in ways that are biologically and socially self-sustaining and mutually supportive of life systems. These are the goals for shaping a new landscape based on fundamental environmental values.

It WILL be clear that the principles I have suggested in this chapter cannot be seen as discrete ideas. They are part of every successful environment and must be seen as an integrated whole. The following two case studies both illustrate and summarize how these principles can be applied to design.

The Making of a Memorable Landscape: Harbourfront, Toronto

When Lieutenant-Governor Simcoe set foot on the shores of Lake Ontario in 1793, it was Toronto's naturally sheltered bay that persuaded him to establish the town and garrison of the future city. In the early years of the nineteenth century, settlement expanded and most of the city's contact with the outside world was by boat. The waterfront was Toronto's gateway and commercial focal point. Wharves were extended into the lake to accommodate the shipping that carried grain, lumber, and other raw materials from the Canadian interior to Britain. With the

coming of the railways in the mid-nineteenth century, commercial shipping increased and the railways effectively isolated the waterfront from the town. In 1912 the newly created Toronto Harbour Commission began a comprehensive waterfront-development plan that over some forty years of implementation included extensive land filling, improvements to the harbor, and the creation of new industrial land. Industrial and port facilities were built. A railway viaduct was constructed again to allow access to the waterfront.

By the 1930s the need for improved vehicular access to the city led to the construction of major arterials and expressway links, introducing additional barriers to the waterfront. With the decline of industrial and commercial shipping in the 1970s, the rehabilitation of the waterfront for recreation and leisure began in earnest, marked by the acquisition of eighty-six acres of waterfront dockland by the federal government as public park.

The Harbourfront Corporation, set up to oversee the new development, began with a promise to the federal government who owned the land, to make Harbourfront financially self-sufficient. Initially, activities and programs were developed to attract people to the waterfront while keeping development costs low. They included such things as a flea market, theater, art gallery, a junk playground. As the waterfront became popular as a place to visit, commercial and residential developments were built, with funding for public facilities coming from revenues from private development. Development, however, began to get out of control when condominium apartments of dubious architectural merit were built twice as high as recommended in the planning framework. Open space began to be squeezed out in favor of private investment. A public outcry ensued that Toronto's last links with its lake were being squandered, that the opportunities for bringing the waterfront back to the people were being lost to private interests, and that Harbourfront was turning into a continuation of the faceless commercial developments of downtown. In 1987, as a result of the continuing criticism of the direction this public waterfront was taking, a design review was launched to guide its future development.[17]

The factors affecting the identity of the waterfront became a crucial aspect of the review. It was apparent that development was ignoring its geographic, historic, and social legacy and turning Harbourfront into another placeless environment of high-rise buildings and vacant grassland parks. Future planning was founded on development principles that addressed some central questions: What makes a dynamic waterfront? How are its open spaces structured and used? How do they contribute to social and environmental health? And

how are they experienced as places? These questions gave rise to certain principles to guide the waterfront development.

Context. The setting of Harbourfront—its larger context—was determined by its natural and cultural history. Since the nineteenth century, the Toronto Islands that enclose the inner harbor have been the city's prime lakefront recreational area. Today, the legacy of their wooded landscape, open spaces, beaches, and natural shoreline include activities suited to their parklike character— swimming, ethnic picnics, boating, funfair, nature education. All these activities are made special by the experience of the ferry that takes people from a dense and noisy downtown to a quiet pastoral landscape of repose, lake breezes, and clear air. The development of the inner harbor as a historic working port has left a legacy of built form, hard edges, shipping docks, industrial grain silos, and railways to the waterfront against which the downtown forms a dramatic urban backdrop.

Opened in 1971, Ontario Place was the first cultural parkland development to initiate the new waterfront revival by bringing people back to celebrate the water's edge and lake, with cultural events, music, open-air amphitheater, shops, and restaurants. The Leslie Street Spit, in contrast, has evolved fortuitously into an urban wilderness and wildlife sanctuary of

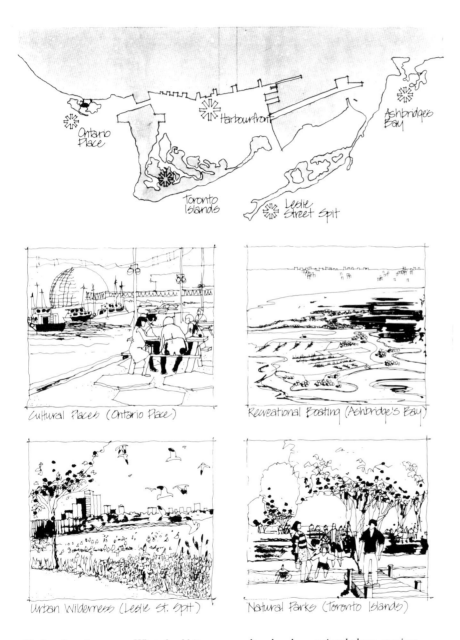

Cultural Places (Ontario Place)

Recreational Boating (Ashbridge's Bay)

Urban Wilderness (Leslie St. Spit)

Natural Parks (Toronto Islands)

Harbourfront in context. What should its essential nature be? The cultural and natural history of Toronto's waterfront has left an astonishingly varied legacy that includes a working port, pastoral leisure landscapes, *cultural and recreational places, marinas, and an urban wildlife wilderness. The basic principle was to accentuate these differences, not undermine them. (Design Panel for Harbourfront 2000)*

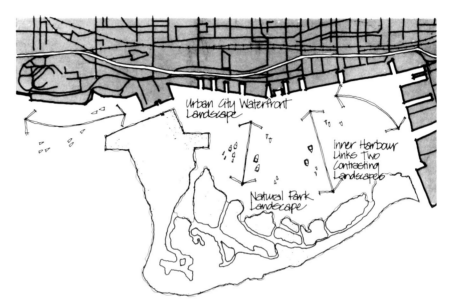

What makes the waterfront special? The urbanized shoreline of the old port, with its docks, warehouses, and industrial buildings, and the wooded pastoral landscape and soft edges of the Toronto Islands, give Toronto's central waterfront a powerful image as a memorable place. The relationship between these two contrasting landscapes, separated by and enclosing the inner harbor, provides the key to its regional identity. (Design Panel for Harbourfront 2000)

great diversity from abandoned efforts to create a new inner harbor in the early 1960s. It is a place to get away from the crowds and urban tensions, to study evolving natural processes, to walk, cycle, and remain quiet while the gulls and the terns make the noise.

Each of these places on the waterfront has its own distinctive landscape, appeal, and social value. The relationship among them, the geographic setting of diverse landscapes, and the harbor that unites them create a powerful image and unique regional identity. These are what make the city memorable. The design principles for Harbourfront were inspired by this setting. Its open spaces were unique in their overall context and very different from other places in the city.

Work and Historical Continuity. The best waterfronts are working ones as well as places for leisure. Toronto is memorable for the vast and dramatic scale of its cargo ships and cranes,

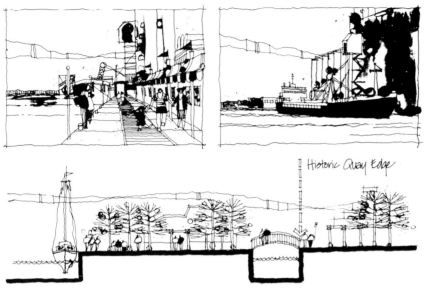

Historic Quay Edge

Maintaining a working historic waterfront. The most interesting waterfronts are those that include work as well as leisure. The vast scale and drama of large cargo boats with their cranes, equipment, quays, sheds, and railways are what make such places so exciting. It is important to incorporate parts of this historic legacy into new planning. (Design Panel for Harbourfront 2000)

In many ways the most interesting parts of the present-day waterfront are the industrial areas where the lakers still dock and unload, and where there is a wealth of unstructured environments and places to be discovered.

for industrial architecture that over-powers people and buildings and an-nounces the lake and waterfront at the end of streets. These are, in many ways, the most interesting parts of present day Toronto, where lakers still dock, load and unload goods, and where there is a wealth of un-structured places waiting to be dis-covered. But the city's days as an industrial port were numbered, so it was necessary to think about work in other ways. Today's working land-scape is one of new forms. The har-bor police, the ferries that ply the inner harbor, the Great Lakes Re-search vessels that dock in the harbor provided alternatives to recreational boats. In addition, the abandoned lakers awaiting the scrap heap, the grain silos splendid in their scale, simplicity, and nineteenth-century functionalism, the rail lines that used to link the harbor with the continent are today grist for adaptive re-use to modern functions while maintaining connections with the past.

Environmental Awareness. Among the many things we do and see on a waterfront, what tells us about where we are and about its connections with the larger environment? There should be opportunities to enhance our awareness of the place, to connect us to it in ways that are educational, instructive, and part of its every-day activities and enjoyment. The educational themes established for Harbourfront link the place to Lake Ontario and the Great Lakes as an ecosystem; to pollution issues and in-tegrated ecosystem management; to the aquatic and terrestrial habitats that unite the system and the cities that depend on it; to the cultural his-tory of exploration, settlement, and sailing; to commerce and urban de-velopment. These themes can occur in a variety of contexts. Among the most appropriate is a rehabilitated lake ship, fitted out as a museum of natural and cultural history and as a special kind of water park for educa-tion, exploration, and adventure.

Climate and Livability. Toronto has a regional and local climate where the cold of winter is as long as the heat of summer. The onshore winds that blow across the waterfront make for pleasant conditions in summer but unpleasant ones in winter. Increasing residential development will lead to year round use of the waterfront and at the same time require protected winter places. Buildings and massed coniferous vegetation are placed to deflect winter winds; deciduous vege-tation shades spaces against the sum-mer sun; arcades and colonnades provide pedestrian comfort at all sea-sons; prevailing off-lake winds are harnessed into works of environmen-tal art, all of which enhance the spe-cial character of the place.

Places for Everyone. Successful public places draw upon residents and visi-tors with uses that offer something to

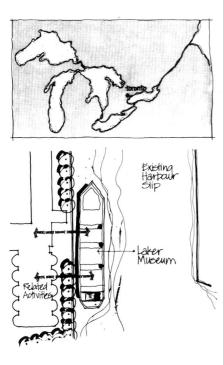

Creating environmental awareness. Harbourfront should encourage and enhance an understanding of Toronto's place on the Great Lakes system in ways that are holistic, instructive, and enjoyable for people of all ages. A Great Lakes museum could combine educational themes: the Great Lakes as an ecosystem; pollution and ecosystem management; aquatic habitats; birds, animals, and fish; the cultural history of sailing, exploration, or settlement. The museum might, appropriately, be incorporated into a rehabilitated laker, moored permanently to a pier. (Design Panel for Harbourfront 2000)

every interest. Open spaces of many kinds reflect this variety including busy plazas, relaxed places, thematic gardens, places for eating, browsing around stores, looking at boats and water—all designed with children as well as adults in mind.

Linkages and Coherence. Toronto's flat topography inhibits the strong visual connections with water that are characteristic of hillside cities such as San Francisco. In one sense, however, the views from Toronto's highrise, downtown core make up for the lack of topography. At ground level there are other opportunities in the alignment of the north-south

street grid, and gentle slopes to the shoreline focus attention on the industrial symbols of the waterfront, the distinctive forms of cranes and ship masts. These north-south links with the city are accentuated with additional landmarks to be seen from a distance. From these general principles, three specific objectives evolved that included the creation of a high-quality, diverse, public, urban, open-space system; a grand pedestrian boulevard linking the entire waterfront; and a distinctive architecture for Harbourfront which would reflect its special character while providing shelter, public activity, and visual focal points along its length.

Overall view of the Harbourfront 2000 model, looking east. Elements of the plan include: public facilities, housing, historic industrial buildings, and open spaces dispersed along the old waterfront quays; the continuous promenade and link; office and more residential building on the north side, with the city in the background. A continuous canopy of deciduous and coniferous trees together with quayside buildings provide year-round climate protection from lake winds and summer sun. (Harbourfront Corporation)

The city on the water. A view of Harbourfront looking north toward the city center. (Harbourfront Corporation)

Harbourfront at Queen's Quay, facing the inner harbor, with the Toronto Islands in the background. On the left, Queen's Quay Terminal, an old industrial port building now renovated for office, shops, and residential apartments on upper floors. (Harbourfront Corporation)

The plan, designed to take Harbourfront into the future, was based on the overall idea of extending the city to the water. Its public spaces would reflect an urban place with all the richness, diversity, life, and ambience, that the special nature of the location affords. One of the great assets of any dynamic urban place is the fact that it represents a mix of design quality, from the good to the bad, from the humdrum to the beautiful, a sense of many interests influencing the form of the city. This is what cities are about and what lends them excitement and diversity.

Cultural Identity and Northern Settlement: Tuvvik

The native villages and towns in northern Canada all bear the marks of recent settlement that has been planned by a remote southern bureaucracy. For the traditionally nomadic Inuit, the concept of permanent community living is a new one that has been imposed, along with southern education, massive unemployment, welfare checks and southern building, by government. It is clear from a visit to Iqaluit,[18] the largest town in the Baffin region with a population of about 3,000 people, that there is little investment in the place. The town, like most others in the north, has the appearance of a

dump. Scrap metal, wooden crates, oil drums, tires, bottles, and tin cans litter the town, creating an eyesore within its powerful and beautiful setting of hills and coastline. The houses, hotels, stores, and other buildings are widely scattered over the bare windblown slopes. They do not feel connected to the landscape. There are few improvements to the immediate surroundings, or demarcations of private space. The wind whistles around each building, and the ground plane moves from street to house with no apparent break except for the occasional ditch. The town's plan, designed for the automobile, is based on a major ring road. Most people, however, get around by skidoo or on foot, making their own, often perilous and unplanned connections between housing, Hudson Bay store, church, community center, or hospital. Unemployment accounts for one third of the Inuit population; alcoholism is rampant; and the sense of identity of a native people eroded.

This loss of identity is a consequence of a paternalistic relationship that has evolved from a white culture whose government controls their lives. Separated from the land, the Inuit lose their own culture and all the coping mechanisms necessary for survival on it.[19] White education is seen as foreign and antithetical to their way of life. Few do well at school and in the words of one Inuit, "education takes the fun out of being

Iqaluit, Baffin Island. There is little investment in the place, or recognition of climate and livability in the layout of northern towns.

The northern settlement as dumping ground: a depressingly similar sight everywhere, with wasted resources that could be used to create local industries.

stupid." [20] As Robert Bechtel has pointed out, the underlying causes of alcoholism lie in this "paternalism-dependency relationship" that has left native people without a sense of life purpose or direction, caught between a native culture that has changed and a white culture to which they do not relate. [21] The sense of cultural purpose and identity will flourish only if given time and encouragement. In addition, the basic problem of living in permanent communities is something that has to be learned by a native people whose culture has been nomadic.

It was into this physical and social context that a treatment center was commissioned to deal with one of the north's most serious problems. Its underlying philosophy was that the "essence of prevention is the empowerment of the Inuit to take control over their own lives." [22] It is a process of socialization where one adult is assigned to another in the role of a brother to assist in his recovery. In Inuit culture he becomes a part of the family, a relationship, termed *anaagisagh* that often becomes more meaningful than that of blood family.

The design process that established the program and architectural requirements for the center became the basis for this re-establishment of local control by the native people. They became an essential part of building and site planning decisions. The Elders expressed a definite pref-

erence for a site overlooking the bay, one that was screened from the town and had access to the land and, above all, had uninterrupted space around the building. In the absence of any plant cover higher than a few centimeters, topography was a major climatic consideration in choosing the site. It was important, therefore, to protect the building in the lee of high ground and away from north winds and snow accumulations.

The Elders said they would like to have a seven-sided building. This raised some interesting issues. It implied a symmetrical form consistent with traditional snow houses that require a location on flat ground. It also implied a large interior space that would remain undivided and open. Traditionally, the Inuit have preferred a large single room for multiple group activities that occur in various parts of the same space at the same time. The Elders raised other points.

There was a need for recreation with a focus on family treatment such as exercise, music, crafts, and a pool table. They wanted a workshop in the building, and space outside for stone, wood, and bone carving, drawing, jewelry making and related crafts, that were seen to be part of the treatment process. They asked for a place to keep hunting equipment for expeditions on the bay. Spaces to enable families to be together during treatment were also necessary. Finally, they wanted as few rules and regula-

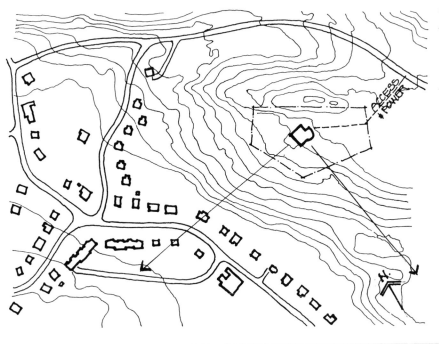

A site separated from the town overlooking the bay was a requirement of the native people for the rehabilitation center.

The siting of the building requires understanding of snow and wind behavior.

tions as possible so that the building would become a home rather than an institution. A main gathering area associated with an outdoor sheltered space was considered important.[23] Other ideas evolved from long hours of observation of the town and its surrounding landscape.

Stone as a Building Material. Building in the town is entirely derived from materials and skills imported from the south and ignores basic considerations of siting, sheltered entrances from wind and snow, privacy and useful outdoor space, hard outdoor surfaces for walking in summer. The local native material is stone. It lies everywhere on the surface of the land and is available in quantity as a potential building material. It has applications for drystone walling for wind-protected spaces and suntraps, and as a paving material for both interior and exterior floors. In addition, in northern settlements recyclable materials such as oil drums are abundant, which, when filled with stone, have potential uses in wall construction.

Stone walls also have the potential

to create sheltered microclimates for growing arctic plants. In the wilderness, wherever a slight rise in the land provides protection from the wind, a southern orientation, and a place for snow to accumulate, plants will grow higher and more vigorously. Applied in this way, walls can be used to bring native vegetation back to a town environment from which they have long disappeared. Links between land, people, and building developed in ways that are useful and practical could, therefore, create a new urban vernacular: the form of the human landscape arising from its geomorphology; the rock, plants, and buildings themselves derived from the needs of shelter and the requirements of current technology. Most important, however, was the proposal to use the construction of the Center to initiate a local stone industry, developed and managed by the native people in Iqaluit.

Growing Food. The Elders made it plain that their interest in plants was in their edibility rather than in their aesthetics, although exotic plants are frequently part of people's living

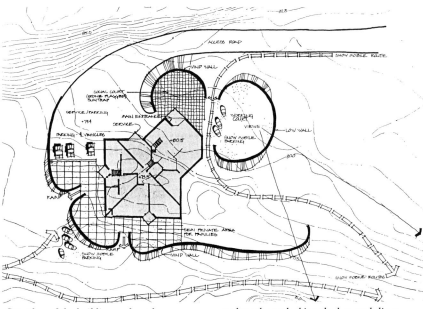

Site plan of the building and outdoor areas, showing the main arrival areas, common interior space associated with outdoor social and working courts, and family treatment rooms. The building has views to the west and south overlooking the bay and direct access to the larger landscape for hunting expeditions. (From Peter Goering; Hough, Stansbury + Woodland)

The native stone is granite, a material that could become the basis for a local building industry run by the native people.

spaces. Locally produced summer leaf vegetables and other nutritious crops grown under protective glass could be practically introduced and vary local diet.

Designer Training Program. Participation in a program to begin training native people as future designers for their own communities has the potential for producing untold benefits for the community at large.

The application of the ideas used in the development of the rehabilitation program can be extended to the town itself, to create local professional and construction skills, horticultural programs for arctic flora propagation for export, and similar northern business enterprises that are relevant to the region and its people.

OVER TIME it may be that the opportunities that come from this initiative will gather momentum and help create a new sense of identity among the native people, one that is tied to the land and to a new form of northern urban living that invests in the environment and the community. The design proposals made are modest, small in scale, and avoid comprehensive or fixed long-range plans. They can only be realized when the native people themselves believe that they make practical sense and can exert control over their development.

There is often an overwhelming temptation to see quick, concrete results in the absence of the motivating local forces that can make things happen. Design in this sense follows the patterns of authority—the superimposed planning that has created the loss of cultural identity and links to the land among native peoples. The true role of design is to sow the seeds from which local processes take off by themselves—doing as little as possible for maximum benefits. It is, therefore, of a very different order from that which is imposed from above as design form. For most of us this is a difficult lesson to learn. But without this understanding of process we ignore the very basis on which regional identity can flourish.

Some Final Reflections

Armed with a broad environmental perspective on the nature of the regional imperative, design can begin to make a contribution to establishing a viable contemporary landscape. It is a perspective that is rooted in ecological and cultural diversity. If we look for it, the inherent potential for diversity shines like a beacon through the placeless dreariness of much contemporary urbanization. The making of memorable places involves principles of evolving natural process and change over time. It involves economy of means where often the less one does to make purposeful change the better. It involves variety and choice that evolve naturally through countless interactions between people and nature, providing a secure basis for ecological and social health. It also has to do with understanding the nature of places as a precursor to making purposeful change, which is a far more significant act of creativity than imposing pre-packaged solutions on the land. The familiar and overworked analogy of the Eskimo carver who, staring at the stone in his hand, wonders what it is within that wants to come out, serves to encapsulate the underlying philosophy of what place is all about. When the carver recognizes what it is, he simply carves the stone to release it. Where the processes of nature are allowed to become part of an organic as opposed to a fixed view of the planning process, the opportunities for regional identity are enhanced. Both human and nonhuman nature are fueled by similar underlying processes and motivating forces. The true nature of the regional imperative has little to do with mega-projects or utopian dreams. It has to do with what is, with understanding the forces that make change and making the most of opportunities wherever and in whatever form they may arise. Eutopia (*good place*) not Utopia (*no place*) is the goal toward which we must strive.

Notes

Introduction

1. John Brinckerhoff Jackson, *Discovering the Vernacular Landscape*, New Haven: Yale University Press, 1984.
2. Eugene P. Odum, "The Strategy of Ecosystem Development," *Science* 164 (April 1969).
3. The term *Inuit,* meaning "the people," is how the Canadian Eskimos refer to themselves.

Chapter 1:
The Regional Imperative

1. Allan Fotheringham, "Oh to Be in England and Bored," *Maclean's Magazine,* Dec. 21, 1987.
2. Bernard Rudofsky, *Architecture without Architects*, New York: Museum of Modern Art, 1964.
3. Grady Clay, "The Emerging Places in North America," Public lecture at the University of Guelph, Ontario, 1986.
4. Bernard Rudofsky, *Streets Are for People,* New York: Van Nostrand Reinhold, 1982.
5. Lewis Mumford, *The City in History*, New York: Harcourt, Brace and World, 1961.

Chapter 2: The Native Landscape and Human Perceptions

1. Michael Laurie, "Nature and City Planning in the 19th Century," in *Nature in Cities,* Ian C. Laurie, ed., Chichester: John Wiley and Sons, 1979.
2. John Ruskin, *Modern Painters,* vol. 4, pt. 5, Orpington: George Allen, 1888.
3. George Seddon, "The Landscapes of China," *Landscape Australia,* 2/82, May 1982.

4. Chiang Yee, *The Silent Traveller: A Chinese Artist in Lakeland,* London: London Country Life, 1937.
5. Yi Fu Tuan, *Topophilia: A Study of Environmental Perceptions, Attitudes and Values,* Englewood Cliffs, N.J.: Prentice Hall, 1974.
6. Neil Evernden, "Beauty and Nothingness: Prairie as Failed Resource," *Landscape* 27, no. 3 (1983).
7. Ibid.
8. Lorus J. Milne and Margery Milne, *The Mountains,* New York: Time Life Books, 1962.
9. J. A. Livingston, *Canada,* Toronto: Natural Science of Canada, Jack McLelland, 1970.
10. Ibid.
11. Nan Fairbrother, *New Lives, New Landscapes,* London: The Architectural Press, 1970.

Chapter 3:
The Cultural Landscape

1. Norman Carver, *Italian Hilltowns,* Kalamazoo, Michigan: Documen Press, 1979.
2. Kevin Lynch, *Managing the Sense of the Region,* Cambridge, Mass.: MIT Press, 1976.
3. Ralf Dahrendorf, *Life Chances,* Chicago: University of Chicago Press, 1979.
4. Frits W. Went, *The Plants,* New York: Time Life Books, 1979.
5. R. V. Davis, *Geology of Cumbria,* Clapham, Lancaster: Dalesman Books, 1979.
6. The term *fells* is the north of England name for hills.
7. James Herriot, *James Herriot's Yorkshire,* London: Michael Joseph, 1979.
8. John Woodforde, *Farm Buildings in England and Wales,* London: Routledge and Kegan Paul, 1983.

9. Carver, *Italian Hilltowns.*

10. Mark Girouard, *Cities and People,* New Haven and London: Yale University Press, 1985.

11. Carver, *Italian Hilltowns.*

12. Ibid.

13. Arthur Raistrick, *The Story of the Pennine Walls,* Clapham, Lancaster: Dalesman Publishing, 1946.

14. Ibid.

15. Carver, *Italian Hilltowns.*

16. Brent Boddy, personal communication.

17. Owen R. Scott, "Utilizing History to Establish Cultural and Physical Identity in the Landscape," *Landscape Planning* 6 (Amsterdam: Elsevier Scientific Publishing, 1979).

18. Jeremy Epstein, "Techniques of Desert Reclamation," in *Landscape Design for the Middle East,* Timothy Cochrane and Jane Brown, eds. London: RIBA Publications, 1978.

19. Woodforde, *Farm Buildings.*

20. Jackson, *Discovering the Vernacular Landscape.*

21. Karl A. Wittfogel, "The Hydraulic Civilizations," in *Man's Role in Changing the Face of the Earth,* William L. Thomas, Jr., ed. Chicago: University of Chicago Press, 1956.

22. W. G. Hoskins, *The Making of the English Landscape,* Harmondsworth, Middlesex: Penguin Books, 1970.

23. Ibid.

24. Scott, "Utilizing History."

25. Girouard, *Cities and People.*

26. Ibid.

27. Ibid.

Chapter 4: Utopias

1. Ian Tod and Michael Wheeler, *Utopia,* New York: Harmony Books, 1978.

2. Lewis Mumford, *The Story of Utopias,* New York: Viking Press, 1922.

3. Bertrand Russell, *History of Western Philosophy,* London: George Allen and Unwin, 1946.

4. Mumford, *The Story of Utopias.*

5. Tod and Wheeler, *Utopia.*

6. Ibid.

7. Donald Worster, *Nature's Economy,* New York: Anchor Press/Doubleday, 1979.

8. Tod and Wheeler, *Utopia.*

9. Robert Fishman, *Urban Utopias in the Twentieth Century,* Cambridge, Mass.: MIT Press, 1982.

10. Ibid.

11. James H. Cassedy, "Hygeia: A Mid-Victorian Dream of a City of Health," *Journal of the History of Medicine and Allied Sciences* 17 (April 1962): 217–28.

12. Ibid.

13. Ibid.

14. Fishman, *Urban Utopias in the Twentieth Century.*

15. Frank Lloyd Wright, *Writings and Building,* Edgar Kaufman and Ben Raeburn, eds., Horizon Press, 1960.

16. Fishman, *Urban Utopias in the Twentieth Century.*

17. Ibid.

18. Philip Boardman, *The Worlds of Patrick Geddes,* London: Routledge and Kegan Paul, 1978.

19. Mumford, *The Story of Utopias.*

20. Milan Simecka, "A World with Utopias or without Them?" in *Utopias,* Peter Alexander and Roger Gill, eds., London: Gerald Duckworth, 1984.

21. Robert Fishman, "Utopias in Three Dimensions: The Ideal City and the Origins of Modern Design," in *Utopias,* Peter Alexander and Roger Gill, eds.

22. Jane Jacobs, *The Death and Life of Great American Cities.* New York: Random House, 1961.

23. Alice Coleman, *Utopia on Trial,* London: Hilary Shipman, 1985.

24. Christian Norberg-Schulz, *Genius Loci: Towards a Phenomenology of Architecture,* New York: Rizzoli International Publications, 1979.

25. Coleman, *Utopia on Trial.*

26. Jacobs, *The Death and Life of Great American Cities.*

27. R. W. Newbury, personal communication.

28. R. W. Newbury et al., "The South Indian Lake Impoundment and Churchill River Diversion," *Canadian Journal of Fisheries and Aquatic Sciences* 41 (1984): 548–57.

29. Theodore Roszak, *The Cult of Information,* New York: Pantheon Books, 1986.

30. Crane S. Miller and Richard S. Hyslop, *California: The Geography of Diversity,* Pomona: Mayfield Publishing, 1983.

31. Ibid.

32. Ibid.

33. Ibid.

34. Ibid.

35. John T. Lyle, "Retooling for Green Thumbs," *Los Angeles Times,* Nov. 18, 1984.

36. Ibid.

37. Water Research Team. *Landscape Design Principles for the City of Industry.* Department of Landscape Architecture, California State Polytechnic University, Pomona, Sept. 1986.

38. Ibid.

39. Hugh A. Holub, "Water Conservation and Wealth: A Tale of Two Cities," *Salt River Project "Waterline,"* Spring 1987.

40. Tucson Active Management Area, Arizona Department of Water Resources, *A Water Issues Primer for the Tucson Active Management Area,* 2d ed., Tucson, July 1983.

41. Southern Arizona Water Resources Association, Minutes of Board of Director's meeting, Aug. 12, 1987.

42. George Barr, "CAP: A Vital Part of Basin Management," *Sawara Waterwords* 4, no. 7 (Tucson, July 1985).

43. Remarks by Lewis C. Murphy, Mayor, City of Tucson, to Governor's Committee on Capital Financing, March 20 1985.

44. Barr, "CAP: A Vital Part of Basin Management."

45. Tucson Active Management Area, *A Water Issues Primer for the Tucson Active Management Area.*

46. Remarks by Lewis C. Murphy, March 20, 1985.

47. Holub, "Water Conservation and Wealth."

48. Ibid.

Chapter 5: The Urban Region and the Loss of Identity

1. Joanne Kates, "Indignity to Cheese Offends the French," *Toronto Globe and Mail,* March 18, 1987.

2. Ibid.

3. "The Boom Towns," *Time Magazine,* June 15, 1987.

4. Ibid.

5. Joel Garreau, "Edge Cities," *Landscape Architecture* 78, no. 8 (Dec. 1988): 48–55.

6. Ibid.

7. Environmental Awareness Center, *Focus.* University of Wisconsin, School of Natural Resources, College of Agricultural and Life Sciences, Cooperative Extension Service, July 1984.

8. Ibid.

9. James Krohe, Jr., "Buy Now, Save Later: A Farmland Proposal," *Planning,* Nov. 1986.

10. Grady Clay, "Why Don't We Do It In the Road?" *Planning,* May 1987.

11. Michael Hough, *City Form and Natural Process,* New York: Van Nostrand Reinhold, 1984.

12. Wayne Baerwaldt and Barton Reid, "Re-reading Suburbia," *City Magazine* 8, no. 1 (Winter 1986).

13. Ibid.

14. *Toronto Globe and Mail,* June 22, 1985.

15. Dimitri Dimakopoulos, Architect, Hough Stansbury + Associates, Landscape Architects. *Taikoo Shing Cityplaza, Site B, Hong Kong.* Client: Swire Properties, Ltd., Hong Kong.

16. Professor Charlie Thomsen, University of Manitoba, personal communication.

17. Ibid.

18. Barrie Greenbie, *Spaces: Dimensions of the Human Landscape,* New Haven and London: Yale University Press, 1981.

19. Alvin Toffler, *Future Shock,* Toronto: Bandan Books, 1970.

20. Ibid.

21. Nan Fairbrother, *New Lives New Landscapes.*

22. Dr. Chen Ming, Director, Audio-Visual Research Centre, West China University, personal communication.

23. Kenneth Helphand, "Landscape Perception: Transportation and Travel," lecture series, University of Oregon.

24. Frank Egler, *The Plight of the Right-of-Way Domain, Part 1,* Mt. Kisco, N.Y.: Future Media Services, 1975.

25. Ibid.

26. Richard Scott, "Ecological and Cultural Process as a Basis for Rural Freeway Right-of-Way Management," Major paper in Environmental Studies, York University, Toronto, Dec. 1987.

27. Sylvia Crowe, *The Landscape of Roads,* London: The Architectural Press, 1960.

28. C. Tunnard and B. Pushkarev, "The Paved Ribbon: The Aesthetic of Freeway Design," in *Man Made America: Chaos or Control?* New Haven and London: Yale University Press, 1963.

29. Grady Clay, *Close-Up: How to Read the American City,* New York: Praeger, 1973.

30. Ibid.

31. D. Appleyard, K. Lynch, and J. R. Myer, *The View from the Road,* Cambridge, Mass.: MIT Press, 1964.

32. Scott, "Ecological and Cultural Process."

33. Ibid.

34. Hough, Stansbury + Woodland, *Urban Vegetation Management. Ottawa River Parkway,* Ottawa: National Capital Commission, 1987.

35. Baerwaldt and Reid, "Re-reading Suburbia."

36. Christopher B. Leinberger and Charles Lockwood, "How Business Is Reshaping America," *The Atlantic Monthly,* Oct. 1986.

37. Ibid.

38. Ibid.

39. LA Forum, "Designing New Towns," *Landscape Architecture* 78, no. 8 (Dec. 1988): 66–75.

40. Richard Hedman, with Andrew Jaszewski, *Fundamentals of Urban Design,* Washington D.C.: Planners Press, 1984.

41. City of Scarborough Planning Department, *Urban Design Guidelines,* Scarborough, Ontario: City of Scarborough, June 1987.

42. Michael Hough and Suzanne Barrett, *People and City Landscapes.* Toronto: Conservation Council of Ontario. 1987.

43. Hough, *City Form and Natural Process.*

44. Ibid.

45. Greater London Council, *A Nature Conservation Strategy for London,* Ecology Book No. 4, n.d.

46. Chris Baines, *The Wild Side of Town,* London: BBC Publications and Elm Tree Books, 1986.

47. W. E. Lautenbach, *Land Reclamation Program 1978–1984,* Ontario: Regional Municipality of Sudbury, 1985.

Chapter 6: Industrial Landscapes and Environmental Perceptions

1. Sylvia Crowe, *The Landscape of Power,* London: The Architectural Press, 1958.
2. Robert L. Thayer, Jr., "Beyond Landscape Guilt," *Landscape Architecture Magazine* 74, no. 6, (Nov.-Dec. 1984).
3. Grady Clay, *Close-Up.*
4. Ibid.
5. Nan Fairbrother, *New Lives, New Landscapes.*
6. The New Alchemist Institute, "Modern Agriculture: A Wasteland Technology," *Journal of the New Alchemists,* 1974.
7. Simon Miles, *Towards a Conservation Strategy for Ontario,* Toronto: Conservation Council of Ontario, 1986.
8. James Krohe, Jr., *Buy Now, Save Later.*
9. Mark B. Lapping, "Towards a Working Rural Landscape," in *New England Prospects: Critical Choices in a Time of Change,* C. H. Reidel, ed., Hanover, N.H.: University Press of New England, 1982.
10. Ibid.
11. Roger Miles, *Forestry in the English Landscape,* London: Faber and Faber, 1967.
12. Ibid.
13. ibid.
14. John H. Noyes, *Woodlands, Highways and People,* Cooperative Extension Service, University of Massachusetts, Publication No. 33, Feb. 1969.
15. Crowe, *The Landscape of Power.*
16. Ministry of Agriculture, Fisheries and Food, *Farming UK,* London: HMSO Books, 1987.
17. Ibid.
18. Countryside Commission, *Countryside Commission News,* no. 30 (Jan.-Feb. 1988).
19. Ibid.

20. Economic and Social Research Council, *Interpreting Future Landscapes,* University of East Anglia, Yorkshire Dales National Park, n. d.
21. James Krohe, Jr., *Buy Now, Save Later.*
22. Ibid.
23. Lapping, *Towards a Working Rural Landscape.*
24. John Dudley Scruggs, "Designing the Bluegrass Horse Farm," *Landscape Architecture* 71, no. 3 (May 1981): 377–79.
25. Ibid.
26. Ibid.
27. Horst Schach, "A Bluegrass Stewartship," *Landscape Architecture* 75, no. 2 (March-April 1985): 48–53.
28. Morgan Dix Wheeler, "Horse Farms: A Designer's Approach," *Landscape Architecture* 75, no. 2 (March-April 1985): 69–77.
29. George Goulty, "Landscape Electric," *Landscape Design* (Aug. 1986): 34–37.
30. Ibid.
31. Hough, Stansbury, + Woodland, Petro-Canada Products, Clarkson Refinery, Mississauga, Ontario, ongoing project.
32. R. S. Dorney, "Urban Agriculture: Beyond the Backyard or an Ecologist's Search for a New Agricultural Paradigm," paper for *Greening of the City: An International Symposium,* Toronto, Feb. 1987.
33. Mark Francis, Lisa Cashdan, and Lynn Paxson, *Community Open Spaces,* Washington: Island Press, 1984.
34. Judith Schonbak, "Factory Farm: A Productive Venture by Herman Miller Inc.," *Landscape Architecture,* 73, no. 1 (Jan.-Feb. 1983): 48–53.

Chapter 7: Tourism

1. Statistical report, Caribbean Tourism Research and Development Centre, 1980.

2. Paul F. Wilkinson, "Strategies for Tourism in Island Microstates," *Annals of Tourism Research* 2, no. 16 (1989): 153–77.

3. Ibid.

4. Bryan Johnson, "Bhutan Clamps Down on Tourism," *Toronto Globe and Mail*, Jan. 21, 1988.

5. Mohamed Tangi, "Tourism and the Environment," *Ambrio* 6, no. 6 (1977): 336–41.

6. Fred P. Bosselman, *In the Wake of the Tourist*, Washington, D.C.: The Conservation Foundation, 1978.

7. Clinton Andrews, "Photographs and Notes on Tourism and Deforestation in the Solu Khumbu, Nepal," *Mountain Research and Development* 3 (1983): 182–85.

8. Ibid.

9. Bosselman, *In the Wake of the Tourist*.

10. Ibid.

11. Phil English, "Bloor Street Leap," 2, no. 3 (1985).

12. Ibid.

13. *Toronto Globe and Mail*, travel section, July 27, 1985.

14. Garrett Eckbo, "The Landscape of Tourism," *Landscape* 12, no. 2 (1969).

15. John Betjeman, *John Betjeman's Collected Poems*, London: John Murray, 1958.

16. Barrie B. Greenbie, *Space and Spirit in Modern Japan*, New Haven and London: Yale University Press, 1988.

17. E. Relph, *Place and Placenessness*, London: Pion, 1976.

18. *Toronto Globe and Mail*, travel section, May 9, 1987.

19. James Marston Fitch, "Preservation Requires Tact, Modesty and Honesty among Designers," *Landscape Architecture Quarterly* (May 1976): 276–80.

20. Pierce F. Lewis, "The Future of Our Past: Our Clouded Vision of Historic Preservation," *Pioneer America*, 7, no. 2 (1985): 1–20.

21. Fitch, "Preservation Requires Tact."

22. Rudofsky, *Architecture without Architects*.

23. Gary B. Clarke, "Tourism and the Parks," paper, Heritage for Tomorrow Conference, Banff, Alberta, 1985.

24. Ibid.

25. Northern Ontario Tourist Outfitters Association, "Northern Ontario Tourism Strategy Highlights," *Ontario Conservation News* 14, no. 10 (Sept. 1987).

26. John A. Livingston, *The Fallacy of Wildlife Conservation*, Toronto: McClelland and Stewart, 1981.

27. *Toronto Globe and Mail*, May 28, 1987.

28. Livingston, *The Fallacy of Wildlife Conservation*.

29. Wilkinson, "Strategies for Tourism."

30. Hough, Stansbury and Associates, *Point Pelee National Park Resource Atlas*, prepared for Parks Canada, Toronto, n. d.

31. John Brinckerhoff Jackson, *The Necessity for Ruins*, Amherst, Mass.: University of Massachusetts Press, 1980.

32. Wilkinson, "Strategies for Tourism."

33. David Suzuki, *Toronto Globe and Mail*, Oct. 10, 1987.

34. Michael Hough, "Environmental Education and Forest Management in Ontario, Canada," in *Land Conservation and Development—Examples of Land-Use Planning Projects and Programs*, F. R. Steiner and N. N. van Lier, eds., Amsterdam: Elsevier Science, 1984.

35. Randy Hester, "Subconscious

Landscapes of the Heart," *Places* 2, no. 3 (1985): 10–22.

36. Ibid.

37. Ibid.

38. Lynch, *Managing the Sense of a Region.*

Chapter 8: Principles for Regional Design

1. Arthur J. Cordell, "The Uneasy Eighties: The Transition to an Information Society," *Alternatives* 14, no. 3–4 (1987).

2. Boardman, *The Worlds of Patrick Geddes.*

3. Tony Hiss, "Experiencing Places 1," *New Yorker,* June 22, 1987.

4. City of Toronto Planning Board, *Natural Parklands,* June 1960.

5. Hester, "Subconscious Landscapes of the Heart."

6. Ibid.

7. Don Kerr and Stan Hanson, *Saskatoon: The First Half-Century,* Edmunton: NeWest Press, 1982.

8. E. H. Zeidler, *Toronto Globe and Mail,* Sept. 5, 1987.

9. Scott, "Ecological and Cultural Processes."

10. Lynch, *Managing the Sense of a Region.*

11. William H. Whyte, *The Social Life of Small Urban Spaces,* Washington, D.C.: The Conservation Foundation, 1980.

12. Ibid.

13. Hough, *City Form and Natural Process.*

14. *Toronto Globe and Mail,* April 18, 1987.

15. Ibid.

16. Jane Jacobs, "Guiding Principles for Streets That Work," Energy Probe symposium, "The Streetscape: Planning and Retrofitting As If People Mattered," Toronto, June 1986.

17. Baird Sampson et al., *Harbourfront 2000,* a report to the Futures Committee of Harbourfront, Sept. 1987.

18. Iqaluit is the new name for Frobisher Bay, located on Baffin Island, Canadian Arctic.

19. Robert B. Bechtel, *Feasibility Report, Tuvvik,* phase 1, 1987.

20. Ibid.

21. Ibid.

22. Ibid.

23. The Tuvvik Steering Committee and Peter Goering, Architect, *Feasibility Report, Tuvvik,* Alcohol and Drug Treatment Centre, phase 3, notes on landscape and site by Michael Hough.

Selected Bibliography

Andrews, Clinton. "Photographs and Notes on Tourism and Deforestation in the Solu Khumbu, Nepal." *Mountain Research and Development* 3 (1983): 182–85.

Appleyard, D., K. Lynch, and J. R. Myer. *The View from the Road.* Cambridge, Mass.: MIT Press, 1964.

Ardrey, Robert. *The Territorial Imperative.* New York: Atheneum, 1966.

———. *The Social Contract.* New York: Atheneum, 1970.

Arizona Department of Water Resources. *A Water Issues Primer for the Tucson Active Management Area.* 2d ed. Tucson: Arizona Department of Water Resources, July 1983.

Attenborough, David. *The First Eden.* London: Collins/BBC Books, 1987.

Austin, Richard. *Wild Gardens.* New York: Simon and Schuster, 1986.

Baerwaldt, Wayne, and Barton Reid. "Re-Reading Suburbia" *City Magazine* 8, no. 1 (Winter 1986): 17–28.

Baines, Chris. *The Wild Side of Town.* London: BBC Publications and Elm Tree Books, 1986.

Baird Sampson et al. *Harbourfront 2000.* Toronto: A report to the Futures Committee of Harbourfront, Sept. 1987.

Barr, George. "CAP: A Vital Part of Basin Management." *Sawara Waterworks* 4, no. 7 (Tucson: July 1985): 2.

Berger, John. *Ways of Seeing.* London: BBC and Penguin Books, 1972.

Betjeman, John. *John Betjeman's Collected Poems.* London: John Murray, 1958.

Boardman, Philip. *The Worlds of Patrick Geddes.* London: Routledge and Kegan Paul, 1978.

Boas, Franz. *The Central Eskimo.* Toronto: Coles Publishing, 1888; facsimile ed., 1974.

Bosselman, Fred P. *In the Wake of the Tourist.* Washington D.C.: The Conservation Foundation, 1978.

Carter, George F. *Man and the Land.* New York: Holt Rinehart and Winston. 1964.

Carver, Norman. *Italian Hilltowns.* Kalamazoo, Mich.: Documen Press, 1979.

Cassedy, James H. "Hygeia: A Midvictorian Dream of a City of Health." *Journal of the History of Medicine and Allied Sciences* 17 (April 1962): 217–28.

Chiang Yee. *The Silent Traveller: A Chinese Artist in Lakeland.* London: London Country Life, 1937.

City of Toronto Planning Board. *Natural Parklands.* Toronto: City of Toronto Planning Board, June 1960.

Clarke, Gary B. "Tourism and the Parks." Banff, Alberta: Paper to the Heritage for Tomorrow Conference, 1985.

Clay, Grady. *Close-Up: How to Read the American City.* New York: Praeger Publishers, 1973.

———. "The Emerging Places in North America." Guelph, Ontario: Public lecture at the University of Guelph, 1986.

———. "Why Don't We Do It in the Road?" *Planning* (May 1987): 18–21.

Coleman, Alice. *Utopia on Trial.* London: Hilary Shipman, 1985.

Cordell, Arthur J. "The Uneasy Eighties: The Transition to an Information Society." *Alternatives* 14, no. 3/4 (1987): 4–7.

Countryside Commission. "Countryside Commission News." Cheltenham, Glos.: No. 30 (Jan.-Feb. 1988).

Crowe, Sylvia. *The Landscape of Power.* London: The Architectural Press, 1958.

———. *The Landscape of Roads.* London: The Architectural Press, 1960.

Dahrendorf, Ralf. *Life Chances.* Chicago: University of Chicago Press, 1979.

Davis, R. V. *Geology of Cumbria.* Clapham, North Yorkshire: Dalesman Books, 1979.

Dorney, R. S. "Urban Agriculture: Beyond the Backyard, or an Ecologist's

Search for a New Agricultural Paradigm." Paper for *Greening of the City, An International Symposium,* Toronto (Feb. 1987).

Eckbo, Garrett. "The Landscape of Tourism." *Landscape* 12, no. 2 (1969): 29–31.

Economic and Social Research Council. *Interpreting Future Landscapes,* Yorkshire Dales National Park. University of East Anglia, n.d.

Egler, Frank. *The Plight of the Right-of-Way Domain. Part 1.* Mt. Kisco, N.Y.: Future Media Services, 1975.

Emberson, Peter. "Growth of a Study Center." *Landscape Design* (London: Aug. 1986): 38–40.

Environmental Awareness Center. "Focus." University of Wisconsin: School of Natural Resources, College of Agricultural and Life Sciences, Cooperative Extension Service, July 1984.

Epstein, Jeremy. "Techniques of Desert Reclamation." In *Landscape Design for the Middle East.* Timothy Cochrane and Jane Brown, eds., London: RIBA Publications, 1978.

Evernden, Neil. "Beauty and Nothingness: Prairie as Failed Resource." *Landscape* 27, no. 3 (1983): 1–8.

———. *The Natural Alien.* Toronto: University of Toronto Press, 1985.

Fairbrother, Nan. *New Lives New Landscapes.* London: The Architectural Press, 1970.

Fishman, Robert. *Urban Utopias in the Twentieth Century.* Cambridge, Mass.: MIT Press, 1982.

———. "Utopias in Three Dimensions: The Ideal City and the Origins of Modern Design." In *Utopias.* Peter Alexander and Roger Gill, eds. London: Gerald Duckworth, 1984.

Fitch, James Marston. "Preservation Requires Tact, Modesty and Honesty among Designers." *Landscape Architecture Quarterly* (May 1976): 276–80.

Forman, Richard T. T., and Michel Godron. *Landscape Ecology.* New York: John Wiley and Sons, 1986.

Fowles, John. *The French Lieutenant's Woman.* Boston: Little, Brown, 1969.

———. "The Tree" [published excerpt] "Seeing Nature Whole," *Harpers Magazine* 259 (Nov. 1979): 49–681.

Francis, Mark, Lisa Cashdan, and Lynn Paxson. *Community Open Spaces.* Washington, D.C.: Island Press, 1984.

Garreau, Joel. "Edge Cities." *Landscape Architecture* 78, no. 8 (Dec. 1988): 48–55.

Giedion, S. *Space, Time and Architecture.* Cambridge, Mass.: Harvard University Press, 1952.

Girouard, Mark. *Cities and People.* New Haven and London: Yale University Press, 1985.

Goulty, George. "Landscape Electric." *Landscape Design* (Aug. 1986): 34–37.

Greater London Council. "A Nature Conservation Strategy for London." London: Ecology Book, no. 4, n.d.

Greenbie, Barrie B. *Spaces: Dimensions of the Human Landscape.* New Haven and London: Yale University Press, 1981.

———. *Space and Spirit in Modern Japan.* New Haven and London: Yale University Press, 1988.

Hardy, Thomas. *Tess of the d'Urbervilles.* London: Macmillan, 1974.

Hedman, Richard. *Fundamentals of Urban Design.* Chicago: Planners Press; Washington D.C.: American Planning Association, 1984.

Herriot, James. *James Herriot's Yorkshire.* London: Michael Joseph, 1979.

Hester, Randolph T. Jr. "Community Design: Making the Grassroots Whole." *Built Environment* 13, no. 1 (1985): 45–60.

———. "Subconscious Landscapes of the Heart." *Places* 2, no. 3 (1985): 10–22.

———. "12 Steps to Community Development." *Landscape Architecture* 75, no. 1 (1985): 78–85.

Hiss, Tony. "Experiencing Places 1." *New Yorker,* June 22, 1987, 45–68.

Holub, Hugh. "Water Conservation and Wealth: A Tale of Two Cities." *Salt River Project "Waterline,"* Spring 1987.

Hoskins, W. G. *The Making of the English Landscape.* Harmondsworth, Middlesex: Penguin Books, 1970.

Hough, Michael. *City Form and Natural Process,* New York: Van Nostrand Reinhold, 1984.

———. "Environmental Education and Forest Management in Ontario, Canada." In *Land Conservation and Development: Examples of Land-Use Planning Projects and Programs.* F. R. Steiner and N. N. van Lier, eds. Amsterdam: Elsevier Science Publishers, 1984.

Hough, Stansbury, + Woodland. *Urban Vegetation Management Study.* Ottawa: National Capital Commission, 1987.

Hough, Michael, and Suzanne Barrett. *People and City Landscapes.* Toronto: Conservation Council of Ontario, 1987.

Jackson, John Brinckerhoff. *The Necessity for Ruins.* Amherst, Mass.: University of Massachusetts Press, 1980.

———. *Discovering the Vernacular Landscape.* New Haven and London: Yale University Press, 1984.

Jacobs, Allan. *Looking at Cities.* Cambridge, Mass.: Harvard University Press, 1985.

Jacobs, Jane. *The Death and Life of Great American Cities.* New York: Random House, 1961.

———. "Guiding principles for streets that work." Energy Probe Symposium, *The Streetscape: Planning and Retrofitting as If People Mattered.* Toronto (June 1986).

Kerr, Don, and Stan Hanson. *Saskatoon: The First Half-Century.* Edmonton, NeWest Press, 1982.

Krohe, James, Jr. "Buy Now, Save Later: A Farmland Proposal." *Planning* (Nov. 1986): 12–16.

LA Forum. "Designing New Towns." *Landscape Architecture* 78, no. 8 (Dec. 1988): 66–75.

Lapping, Mark B. "Towards a Working Rural Landscape." In *New England Prospects: Critical Choices in a Time of Change.* C. H. Reidel, ed. Hanover, N.H.: University Press of New England, 1982.

Laurie, Ian C., ed. *Nature in Cities.* Chichester: John Wiley and Sons, 1979.

Laurie, Michael. "Nature and City Planning in the 19th Century." In *Nature in Cities.* Ian C. Laurie, ed. Chichester: John Wiley and Sons, 1979.

Lautenbach, W. E. *Land Reclamation Program 1978–1984.* Regional Municipality of Sudbury, Ontario, 1985.

Leinberger, Christopher B., and Charles Lockwood. "How Business Is Reshaping America." *The Atlantic Monthly,* Oct. 1986, 43–52.

Lewis, Peirce F. "The Future of Our Past: Our Clouded Vision of Historic Preservation." *Pioneer America* 7, no. 2 (1985): 1–20.

Livingston, John A. *Canada.* N.S.L. Natural Science of Canada; Toronto: Jack McLelland, 1970.

———. *The Fallacy of Wildlife Conservation.* Toronto: McClelland and Stewart, 1981.

Lopez, Barry. *Arctic Dreams.* New York: Charles Scribner's Sons, 1986.

Lynch, Kevin. *The Image of the City.* Cambridge, Mass.: MIT Press, 1960.

———. *Site Planning.* Cambridge, Mass.: MIT Press, 1962.

———. *Managing the Sense of the Region.* Cambridge, Mass.: MIT Press, 1976.

McHarg, Ian L. *Design with Nature.* New York: Natural History Press, 1969.

Meinig, D. W., ed. *The Interpretation of*

Ordinary Landscapes. New York: Oxford University Press, 1979.

Meyrowitz, Joshua. *No Sense of Place.* New York: Oxford University Press, 1985.

Miles, Roger. *Forestry in the English Landscape.* London: Faber and Faber, 1967.

Miles, Simon. *Towards a Conservation Strategy for Ontario.* Toronto: Conservation Council of Ontario, 1986.

Miller, Crane S., and Richard S. Hyslop. *California: The Geography of Diversity.* Pomona: Mayfield Publishing, 1983.

Milne, Lorus J., and Margery Milne. *The Mountains.* New York: Time-Life Books, 1962.

Ministry of Agriculture, Fisheries and Food. *Farming UK.* London: HMSO Books, 1987.

Mumford, Lewis. *The Story of Utopias.* New York: Viking Press, 1922.

———. *The City in History.* New York: Harcourt, Brace and World, 1961.

New Alchemist Institute. "Modern Agriculture: A Wasteland Technology." *Journal of the New Alchemists* (1974): 45–55.

Newbury, R. W., et al. "The South Indian Lake Impoundment and Churchill River Diversion." *Canadian Journal of Fisheries and Aquatic Sciences* 41, no. 4 (1984): 548–57.

Norberg-Schulz, Christian. *Genius Loci: Towards a Phenomenology of Architecture.* New York: Rizzoli International Publications, 1979.

Northern Ontario Tourist Outfitters Assocaition. "Northern Ontario Tourism Strategy Highlights." *Conservation News* 14, no. 10 (Ontario: Sept. 1987).

Noyes, John H. *Woodlands, Highways and People.* Cooperative Extension Service, University of Massachusetts, Publication No. 33, Feb. 1969.

Odum, Eugene P. "The Strategy of Ecosystem Development." *Science,* 164 (April 1969).

Open University, in association with the

Countryside Commission. *The Changing Countryside.* London: Christopher Helm: 1985.

Pressman, Norman, ed. *Reshaping Winter Cities.* Waterloo, Ontario: Published under the auspices of Livable Winter City Association, University of Waterloo Press, 1985.

Raistrick, Arthur. *The Story of the Pennine Walls.* Clapham, North Yorkshire: Dalesman Publishing, 1946.

Relph, E. *Place and Placenessness.* London: Pion, 1976.

Roszak, Theodore. *The Cult of Information.* New York: Pantheon Books, 1986.

Rudofsky, Bernard. *Architecture without Architects.* New York: Museum of Modern Art, 1964.

———. *Streets for People.* New York: Van Nostrand Reinhold, 1982.

Ruskin, John. *Modern Painters.* Vol. 4, Pt. 5. Orpington: George Allen, 1888.

Russell, Bertrand. *History of Western Philosophy.* London: George Allen and Unwin, 1946.

Schach, Horst. "A Bluegrass Stewartship." *Landscape Architecture* 75, no. 2 (March-April 1985): 75–76.

Schonbak, Judith. "Factory Farm: A Productive Venture by Herman Miller Inc." *Landscape Architecture* 73, no. 1 (Jan.-Feb. 1983): 48–53.

Scott, Owen R. "Utilizing History to Establish Cultural and Physical Identity in the Landscape." *Landscape Planning* 6 (1979): 179–203.

Scott, Richard. "Ecological and Cultural Process as a Basis for Rural Freeway Right-of-Way Management." Toronto: Major paper in Environmental Studies. York University (Dec. 1987).

Scruggs, John Dudley. "Designing the Bluegrass Horse Farm." *Landscape Architecture* 71 (May 1981): 377–79.

Seddon, George. "The Landscapes of China." *Landscape Australia,* 2/82 (May 1982): 116–32.

Simecka, Milan. "A World with Utopias

or without Them?" In *Utopias.* Peter
Alexander and Roger Gill, eds. London: Gerald Duckworth, 1984.

Smith, Kidder. *Switzerland Builds.* London: The Architectural Press, 1950.

Tangi, Mohammed. "Tourism and the Environment." *Ambrio* 6, no. 6 (1977): 336–41.

Thayer, Robert L., Jr. "Beyond Landscape Guilt." *Landscape Architecture* 74, no. 6 (Nov.-Dec. 1984): 48–55.

Time Magazine. "The Boom Towns." June 15, 1987, 14–19.

Tod, Ian, and Michael Wheeler. *Utopia.* New York: Harmony Books, 1978.

Toffler, Alvin. *Future Shock.* Toronto: Bandan Books, 1970.

Tuan, Yi-Fu. *Topophilia: A Study of Environmental Perceptions, Attitudes and Values.* Englewood Cliffs, N.J.: Prentice Hall, 1974.

Tunnard, C., and B. Pushkarev. "The Paved Ribbon: The Aesthetic of Freeway Design." In *Man Made America: Chaos or Control?* New Haven and London: Yale University Press, 1963.

Water Research Team. *Landscape Design Principles for the City of Industry.* Pomona: California State Polytechnic University, Department of Landscape Architecture. Sept. 1986.

Went, Frits W. *The Plants.* New York: Time-Life Books, 1979.

Wheelock, Morgan Dix. "Horse Farms: A Designer's Approach." *Landscape Architecture* 75, no. 2 (March-April 1985): 69–77.

Whyte, William H. *The Social Life of Small Urban Spaces,* Washington D.C.: The Conservation Foundation, 1980.

Wilkinson, Paul F. "Strategies for Tourism in Island Microstates." *Annals of Tourism Research* 16, no. 2 (1989): 153–77.

Wittfogel, Karl A. "The Hydraulic Civilizations." In *Man's Role in Changing the Face of the Earth.* William L. Thomas Jr., ed. Chicago: University of Chicago Press, 1956.

Woodforde, John. *Farm Buildings in England and Wales.* London: Routledge and Kegan Paul, 1983.

Worster, Donald. *Nature's Economy.* New York: Anchor Press/Doubleday, 1979.

Wright, Frank Lloyd. *Writings and Building.* Edgar Kaufmann and Ben Raeburn, eds. Horizon Press, 1960.

Yee, Chiang. *The Silent Traveller: A Chinese Artist in Lakeland.* London: London Country Life, 1937.

———. *The Silent Traveller in Edinburgh.* London: Methuen, 1948.

Index